Pediatric Molecular Pathology: Quantitation and Applications

Perspectives in Pediatric Pathology

Vol. 16

Series Editors

Harvey S. Rosenberg
Professor of Pathology and Pediatrics,
University of Texas Medical School, Houston, Tex.

Jay Bernstein
Director of Anatomic Pathology,
William Beaumont Hospital, Royal Oak, Mich.

KARGER

Basel · Freiburg · Paris · London · New York · New Delhi · Bangkok · Singapore · Tokyo · Sydney

Pediatric Molecular Pathology: Quantitation and Applications

Perspectives in Pediatric Pathology

Vol. 16

Series Editors

Harvey S. Rosenberg
Professor of Pathology and Pediatrics,
University of Texas Medical School, Houston, Tex.

Jay Bernstein
Director of Anatomic Pathology,
William Beaumont Hospital, Royal Oak, Mich.

KARGER

Basel · Freiburg · Paris · London · NewYork · New Delhi · Bangkok · Singapore · Tokyo · Sydney

Pediatric Molecular Pathology: Quantitation and Applications

Volume Editors

A. Julian Garvin, Charleston, S.C.
Timothy J. O'Leary, Washington, D.C.
Jay Bernstein, Royal Oak, Mich.
Harvey S. Rosenberg, Houston, Tex.

29 figures, 2 color plates and 10 tables, 1992

KARGER

Basel · Freiburg · Paris · London · NewYork · New Delhi · Bangkok · Singapore · Tokyo · Sydney

Perspectives in Pediatric Pathology

Library of Congress Card No. 72-88828

Bibliographic Indices
   This publication is listed in bibliographic services, including Current Contents® and Index Medicus.

Contents

Foreword

Most pathologists would agree that immoderate love of morphology has meant both boon and bane in their professional practice. Until recently, its overall effects would still be tallied on the credit side: morphologic studies, despite their limitations, were urgently needed, zealously procured, and universally regarded as foremost in the hierarchy of diagnostic investigations. Everyone knew that merely 'to see' is utter lack of investigative sophistication; thus, this method could be predicted to be uninformative. But the morphologically-inclined pathologist could counter that no better alternative existed; and that the record showed astounding conceptual feats achieved thanks to seemingly paltry methods. Then, talking to himself, the pathologist would say that sundry non-morphologic techniques had come and gone before that purported to replace histopathology, and every time the study of structure retained its privileged status as the irreplaceable foundation on which all other medical workup accrued. After all, the entire edifice of modern medicine rested upon the keystone of structure-function correlation, and it would be a long time before structural analysis – especially of tumors – could be supplanted. The obligatory corollary to this reassuring thought was to continue the byzantine subtyping of morphological appearances. The pathologist's duty was, in Rosenberg's [1] felicitous *bon mot,* 'to tabulate, classify, and replace clinical clutter with morphological clutter, but in neat bins'.

Proof of the unprecedented power of recent technical innovations is their ability, realized or potential, to wrest the primacy from the morphologic approach to diagnosis, and in some cases to supplant it altogether. And it is fitting that the present issue of *Perspectives in Pediatric Pathology,* like the previous one, should devote a generous space to the dissemination and review of such methods as portend a genuine re-evaluation of a cherished tradition.

In the following pages, Dr. Maria Tsokos recapitulates previous efforts to dispose neuroectodermal neoplasms into 'neat bins'. This is a field threatened by imminent chaos, populated as it is by tumors formerly assumed to be not neural crest-derived, but which may sometimes assume a neurogenic phenotype (i.e., Ewing's); tumors of diverse biology that are morphologically indistinguishable from each other; tumors that may show various differentiations simultaneously ('polyhistioma'), or no differentiation at all. To extricate oneself from this labyrinth, one must have Ariadne's thread in the form of clear diagnostic criteria or a map in the form of an all-encompassing conceptual scheme. Dr. Tsokos's scholarly monograph (nearly 400 references) attempts to provide just such guiding conceptual thread. No comparable comprehensive review, prepared from the vantage point of the pediatric pathologist, is available. Although it does not exorcise entirely the specter of 'round cell tumors', it can only be read with profit. Her monographic article emphasizes the need to correlate morphologic findings with data from flow cytometry, cytogenetics, and molecular biology. That this is no idle commonplace was brought home by a case of our recent experience.

A newborn infant was prenatally diagnosed with an intra-abdominal tumor. By age (congenital), location (adrenal), clinical or radiological data (hepatic metastases, speckled calcification), and clinical laboratory tests (mildly elevated catecholamines), the presumptive diagnosis was neuroblastoma. At birth, the decision to needle-biopsy was made. Because needle biopsies are not without risk, and given the uncertainty of procurement of tissue at every needle thrust, a determination had to be made as to the best ways to utilize whatever tissue was obtained. There is little question that until recently the first priority would have gone to morphology, since 'the diagnosis comes first'. In the present state of progress, however, the priorities were different. Tissue had to be secured first for ascertainment of DNA ploidy and N-myc copy number; if any extra material was available, it could go for histopathological processing. This novel distribution of priorities reflects clinicians' contemporary perceptions. Useful information was deemed likelier to issue from study of nonmorphological parameters. It is, of course, theoretically possible that the mass might have been 'something else', i.e., not a neuroblastoma. But in the current state of advancement this probability was judged so low as not to modify the planned clinical conduct. As it turned out, sufficient tissue was obtained for all purposes, the diagnosis was neuroblastoma, and the nonmorphologic parameters established its biologically quiescent nature (aneuploidy, unamplified N-myc). The patient was not treated, and is evolving favorably.

This example vividly portrays the ability of new tools to potentially displace (although not, in a strict sense, to replace) morphologic techniques to a secondary position wherein they are no longer deemed the chief determinants of clinical conduct. It is a highly selected example, but the power of molecular biology tools is likely soon to shake up received opinion in other instances. A pertinent example is the ingenious polymerase chain reaction (PCR), a technique discussed in this issue by Dr. Beverly Rogers. PCR amplifies DNA and RNA sequences in such a way that, starting with incredibly minute targets, of the order of magnitude of a few molecules, a final product is obtained on which a diagnosis can be based. A little over a year ago, 'bacillary angiomatosis' [2] was shown to be caused by a rickettsia-like organism; the investigators used primers complementary to ribosomal RNA genes proper to eubacteria. Because DNA sequences are unique to the different branches of the evolutionary tree of bacteria, appropriate classification of the organism was possible. Thus, a diagnosis of bacterial disease can be made by PCR even when the agent is not visualized in tissue sections, and does not grow in cultures. This technique is applicable to archival material available to pathologists, as shown in Dr. Rogers' review. Hence, pathologists using PCR have an exquisitely sensitive tool to detect pathogens, minor structural changes and rearrangements in genes, residual tumor cells, and a variety of other abnormalities of interest to oncology, virology, microbiology, genetics and forensic medicine.

Lest we be led to the wrong conclusion that morphology's potential has been exhasuted, the interesting communication by Drs. Karen Schmidt and Carlo Pesce restores our confidence in its bright new vistas. Morphometrics, heretofore largely confined to research, is shown to hold exciting promises in diagnosis. Densitometry and computer-assisted image analysis are technical advances that do away with the dreary counting routines, or the torture-like serial sectionings of the past. Schmidt and Pesce's presentation is lucid, and devoid of the constant references to mathematical principles that tend to intimidate the non-mathematical reader. Extended discussion of applications to renal pathology is especially felicitous, because the pathology of renal biopsies has been highly systematized, and its attending procedures are generally standardized. Areas of study with high potential for meaningful results are clearly laid out. Pediatric pathologists are concerned with disease in growing organisms. Refinements in quantitative methods of practical implementation ought to have special appeal for them.

References

1   Rosenberg HS: Pediatric pathology in the time of AIDS; in Garvin AJ, O'Leary TJ,
    Bernstein J, Rosenberg HS (eds): Pediatric Molecular Pathology. Perspect Pediatr
    Pathol. Basel, Karger, 1991, vol 15, p 12.
2   Relman DA, Loutit JS, Schmidt TM, Falkow S, Tompkins LS: The agent of bacillary
    angiomatosis. An approach to the identification of uncultured pathogens. N Engl J
    Med 1990;323:1573–1580.

F. Gonzalez-Crussi, MD, Children's Memorial Hospital, Chicago, IL 60614 (USA)

Board of Editors

Garvin AJ, O'Leary TJ, Bernstein J, Rosenberg HS (eds): Pediatric Molecular Pathology: Quantitation and Applications. Perspect Pediatr Pathol. Basel, Karger, 1992, vol 16, pp 1–6

Founders of Pediatric Pathology:
John Emery (1915–)

A.H. Cameron

Honorary Senior Research Fellow, The Children's Hospital, Birmingham, UK

John Emery

John Emery's work with cot-deaths over many years has gained him an international reputation. Because one of his own children died in infancy, his scientific investigations have always accompanied a compassionate understanding of the problems facing young parents. His goal has always been prevention, with pathological studies providing an important aspect of community paediatrics leading him to extend his activities beyond the laboratory and into close cooperation with the Health Visitors. He regards the Health Visitors as a vital arm of the child nursing services because they do not restrict their attention to just those infants who are brought to the well-baby clinics or who are sick enough to be referred to hospital. The result has been a most dramatic and rewarding fall in infant deaths in Sheffield that has set an example for the other cities in the UK and elsewhere. Can any pathologist claim to greater effectiveness?

He employed two methods as the basis of his work on cot-deaths. First was the meticulous morphological post mortem study, accompanied by statistically controlled comparisons with hospital deaths. Much of his published work considered one organ at a time, for example, the lymphoid aggregates in the lung or the progress of ossification at the costochondral junction. This led inevitably to a study of normal development, which he soon discovered was based on very scanty data at that time. Paradoxically in his early investigations, he planned to use cot-death material as normal controls for other projects, but he soon realized the error. His second method was an investigation of the domestic environment, again with appropriate controls. By this means he identified certain risk factors which allowed the preventative community paediatric services to concentrate their attention on particular families. He always emphasized the dependence on teamwork and, in particular, the close involvement of the Health Visitor service.

Although cot-deaths have been the main topic of his published work, other subjects have not been neglected. There is scarcely any aspect of paediatric pathology that John has not written about, and his papers on lung development, on spina bifida, and on hydrocephalus are of particular interest. His papers are characteristically concise with a clear message, even if only to point out the current state of ignorance. He has been a prolific writer, producing an average of almost one paper per month during his time in Sheffield. Many of his papers have been published in the *Archives of Disease in Childhood;* it is almost impossible to find a volume without an example of his work.

Although his papers have been of great importance, his oral presentations at society meetings have made an even greater contribution to current thoughts on paediatric medicine. His lively and entertaining lectures are always informative, often delivered without script or even notes, usually managing to introduce an original concept or a new way of looking at familiar things. His fondness for describing pathology as a dynamic subject and not just an interesting series of static pictures is coupled with an ability to conjure up illuminating graphic images – such as the resemblance of maturing chondrocytes to a pan of simmering porridge!

His particular attribute of educated originality has much to do with his upbringing. He was born in 1915, the eldest son of a village schoolmaster in the Forest of Dean, a rather isolated part of England near the Bristol Channel. The area has somewhat mystical traditions, its people tend to have independent minds sharpened by the laws of nature. They also tend to be a hardy and fertile folk. John is one of five children, and his father is now an alert 96 years of age, still living in the Forest. John and his wife Mytts (nee Marjorie Mytten) have two daughters and four sons. One of the daughters is a lecturer in a Teacher Training College and the other is an engineer with the Thames Water Authority, the first woman to be appointed to such a post. One son is a general practitioner, one is a trainee obstetrician, one is a hospital administrator, and one is a civil engineer.

John attended the local grammar school at Lydney, and in 1939 he qualified in medicine at Bristol University. During his post-graduate training in paediatrics the Bristol Childrens' Hospital was bombed in an air-raid, and for a time he was in charge when the hospital was evacuated to Weston-super-Mare. During the war he spent a year in pathology, which was then traditional for trainee consultants, and he 'ended up as pathologist in charge of the evacuated hospital laboratories of the Bristol University Teaching Hospital' [J. Emery, pers. commun.]. His laboratory experience at this stage was mainly in biochemistry and haematology, and he gained the MD with a thesis on sugar metabolism in the coeliac syndrome. After the war he found no career in clinical paediatrics, so he decided to take up paediatric pathology. He was appointed as Consultant in Sheffield Childrens' Hosital in 1947. At that time 'the total laboratory services consisted of one small room in Casualty with one technician who did haemoglobins, red cell counts and urine deposits'.

Apart from building up the resources of the department, John entered a field in its infancy. At the time there were very few paediatric pathologists in the world, let alone the UK. He acknowledges guidance from Agnes McGregor in Edinburgh and Henry Barr in Birmingham, but I suspect he was largely self-taught, as many of us were is those days. There was very little known about normal development in early infancy, and his early work concentrated on a comparison between malformations and 'normal' development as seen in cot-deaths. Although he was on an international paediatric tumour panel, he gradually withdrew from this activity because he did not want to 'spend half my life doing reference diagnostic work' [J. Emery, pers. commun., July 1984] and because there seemed to be little point in duplicating the activities of the Childrens' Tumour Registry then being developed in Manchester. He also had a special interest in tuberculosis that receded as the disease was brought under antibiotic control.

As a young man, John had strong pacifist convictions. He became a conscientious objector which he believes prevented his career in clinical paediatrics from developing. Later, he was disillusioned by the pacifist movement and his activities became channeled into international organizations in pathology. He was a member of council in the International Academy of Pathology and one-time President of the European Society of Social Paediatrics.

Fertility is one of John's outstanding attributes, not only the quantity and quality of his family, nor his establishment of a vineyard in the Forest of Dean, but the fertility of his mind. No matter what the topic of discussion, at a professional meeting or social gathering, he usually has some refreshing view or a new idea to put forward. His fertility also extends to the creation of societies, for he is a founder member and indeed a main instigator of the Paediatric Pathology Society (PPS), the International Paediatric Pathology Association (IPPA), the Developmental Pathology Society, the Foundation of the study of Infant Deaths, and the Society for Research in Hydrocephalus and Spina Bifida. He was secretary of the PPS for the first 10 years and twice its president. In the IPPA he was initially chairman of the steering committee and then chairman of council for the first 6 years. Toghether with Johannes Huber he organized the first five annual Advanced International Training Courses in Paediatric Pathology on behalf of the IPPA.

John has had a very active role in regional and national affairs related to pathology and paediatrics. For 17 years he examined in paediatric pathology for the Royal College of Pathologists and was responsible for slanting the examination towards paediatric pathology if the candidate wished. In the Royal Society of Medicine he was president and later editor for the Section of Paediatrics. In Sheffield he was secretary and subsequently chairman of the Childrens' Hospital Medical Staff Committee and chairman of the Centenary

Committee, which has the important function of allocating research money. As some indication of the appreciation for his work, he has been given an honorary fellowship of the British Paediatric Association, the French Paediatric Association, the British Association of Paediatric Surgeons, and the South African Association of Pathologists. He was and still is widely sought after as visiting speaker, and he has given the Still Memorial Lecture to the British Paediatric Association and the Hugh Cairns Memorial Lecture in Australia.

'Since his retirement in 1980 he has remained as busy as ever. One just cannot imagine that he would put up his feet and take things easy; his fertile mind would rebel at the idea. The University Department of Paediatrics has appointed him an Honorary Professor and provided him with suitable accommodation. He continues with research into cot-deaths and remains a member of the Welfare and Research Committees of the Foundation for the Study of Infant Deaths. He has become equally concerned with the promotion of child health. His experience and wisdom are put to very good use in the Sheffield Community Health Council, and he is chairman of the Maternity and Child Services Committee. He pursues similar interests in the Community Paediatric Research Society, of which he has been chairman since its inauguration in 1980.

Great though his achievements are in the field of paediatrics, I suspect he derives as much, and possibly more, pleasure from his other activities, especially those related to the arts. Most people who know him are familiar with the portrait sketches he makes of speakers at scientific meetings. These are necessarily completed against the clock, and it is remarkable that they do not appear to interfere with his subsequent ability to make the most pertinent comment or put the most penetrating question to the speaker. He is a past president of the Sheffield Fine Arts Society and of the Sheffield Museum Society. The city museum and art gallery are fortunately situated in the park immediately opposite the Childrens' Hospital. John is familiar with many art galleries at home and abroad, and a visit with him can amount to a personal guided tour. Decorated cast iron, prehistoric art and art symbolism are among his particular interests and he has lectured fairly extensively on these subjects at home and abroad. John was the first chairman of the Conservation Society in Sheffield. His interests in these fields are shared by his wife, Mytts, who qualified in history at Bristol University.

On top of all this, John has many hobbies including woodwork, sailing and playing the clarinet. He is no mean bard, and his first three publications were poetical. It is no surprise to learn that he is an ex-president of the Sheffield Literary and Poetry Society. In his spare time (!) John continues to cultivate his vineyard, to make his own wine, and to brew mead from his father's honey.

Many are those who have benefited from his works and fortunate are those who have known him.

Dr. A.H. Cameron, 31, Shrublands, Moorend Road, Charlton Kings, GB-Cheltenham, GL53 0NB (UK)

Garvin AJ, O'Leary TJ, Bernstein J, Rosenberg HS (eds): Pediatric Molecular Pathology: Quantitation and Applications. Perspect Pediatr Pathol. Basel, Karger, 1992, vol 16, pp 7–26

Molecular and Histological Analysis of Embryonic Development

David P. Witte, Bruce J. Aronow, Judith A.K. Harmony

Departments of Pathology, Pediatrics, Pharmacology and Cell Biophysics, University of Cincinnati College of Medicine, and Children's Hospital Research Foundation, Cincinnati, Ohio, USA

Introduction

Clarifying the basis of congenital malformations and other embryopathies, which affect as many as 3% of all newborns [1], requires a thorough understanding of the genetic and molecular processes that occur during embryogenesis. During this critical phase in development, several opposing forces are responsible for regulating the expression of the genome. By the differential expression of genes and activities of gene products during development, the diversity of cell types in an organism is established. Each cell within the developing organism must follow the same set of instructions based on a one-dimensional array of genetic information in DNA, but it must interpret them in relation to time and circumstance to produce the complex three-dimensional adult organism [2, 3]. Although approaches to embryology have changed over the years, many of the older issues of developmental biology remain pertinent today and also remain poorly understood. Modern molecular techniques can address old questions regarding morphology, programmed cell death, cellular specialization, organ development, cell growth, tissue induction, and adaptive responses that occur during development.

Embryology began as a descriptive science of the biological processes that regulate and ultimately produce complex biological forms from single cells. In the midst of cellular division, cellular differentiation results from critical cell-type specific gene regulation, RNA splicing, translation, and protein composition. These biochemical changes lead to self-assembly, directed assembly, disassembly, growth and replication, growth arrest and programmed cell death, and the establishment of many metabolic pathways

in the developing organism allowing differentiated cells to establish specialized structures and functions. The study of developmental biology is by
convention defined as the series of progressive, nonrepetitive steps that occur
in the life history of an organism [3]. Many of the instructions for these steps
are encoded in the DNA sequences of the embryonic germ cell, tightly and
orderly regulated in both time and space to develop an organism successfully.
Developmental biologists search for general principles that govern the
interplay between the genetic instructions and the mechanisms forming
complex biological life forms [2, 3].

Classical Embryology

Early experimental work in developmental biology provided detailed
descriptions of preimplantation embryos [4], oviduct transfer [5], and in vitro
observation of dividing morulae [6]. Landmark studies in the late 1920s
showed that a small piece of embryonic tissue could organize an entire second
body axis following transplantation to an ectopic site in a host embryo [7].
These experiments led to the concepts of embryonic induction and organizers.
The search for the chemical basis of these interactions, however, revealed that
a variety of nonspecific agents could mimic the action of the postulated
organizer [8, 9]. These findings suggested little likelihood for identifying
specific endogenous agents responsible for the induction process. Rather, the
concept that a specific chemical inducer was responsible for evoking the
response in developing tissue gave way to the concept that the factors involved
in the developmental process were inherent to the tissue itself.

Experimental Embryology

In vitro culture techniques allowed experimental manipulation of early
embryos [10–13]. The effects of single cell ablation in the developing
blastomere, as well as the development of chimeras, provided important
information about cell lineage in the different embryonic tissues. These
approaches also gave some insight into the time schedule for commitment to
specialized cell types by primitive pluripotent progenitor cells [14–16].
Currently, developmental biologists apply a variety of recombinant DNA
techniques to address issues in understanding the process of development.
The direct introduction of foreign DNA into the blastocele cavity of mouse

blastocysts showed that new genetic material could be integrated into genomic DNA of developing tissue [17]. Eventually the microinjection of cloned genes into mouse embryos allowed the study of specific genes [18–22].

Enzyme Histochemical Analysis of Development

The techniques described above have provided valuable information about organogenesis, cellular derivation, DNA regulation and developmental biological processes. Understanding important cell-cell interactions and the local effects of specific cellular products on developing tissues, however, requires the specific identification and accurate determination of the cellular localization of gene activation or the products of gene translation. The cellular localization of enzyme activities in the developing placenta and embryo represents some of the earliest attempts at understanding these regulated processes during development. In the developing human placenta, alkaline phosphatase activity has been localized in the 13-day conceptus to the syncytiotrophoblast and to the embryonic disc [23]. In the uterus of pregnant mice, 5′-nucleotidase activity has been localized to decidual cells at the time of implantation although most of the decidual cells lost this activity during the later stages of gestation [24]. Adenosine deaminase enzyme activity has also been localized to decidual cells around the implantation chamber early in gestation and to some trophoblastic tissue [25]. Although the high level of activity of these enzymes and their restricted cellular distribution at the implantation site remains unexplained, the studies suggested that these highly induced enzymes were important to implantation and possibly to the survival of the developing fetus. We have recently found in addition to 5′-nucleotidase and adenosine deaminase activity, both of which are part of the purine catabolic pathway, purine nucleoside phosphorylase and xanthine oxidase activities [26]. These enzymes comprise the complete purine catabolic pathway and are coexpressed in the same types of decidual and trophoblastic cells during the gestational period of the mouse. We also showed histochemically that these enzymes were coexpressed within the same cell types and that their activities were functionally linked together, so that potentially toxic purine metabolites could be completely degraded to nontoxic uric acid (fig. 1A, B) The importance of this pathway is suggested by the lethal effect of some specific inhibitors of purine catabolism (e.g. deoxycoformycin) that are lethal to the developing embryo [27]. Although some of these enzymes had been previously detected in the placenta, the determination of the specific cellular distribution provided new understanding about a metabolic pathway that may be important to the implantation process and to the survival of the developing embryo.

Fig. 1. a Histochemical analysis of purine catabolic enzyme in situ showing blue reaction product in the decidual cells immediately adjacent to the embryo indicating high activity of the entire purine catabolic enzyme pathway in a day 6.5 mouse implantation site. *b* Day 6.5 mouse implantation site with embryo in the center surrounded by prominent decidual cells. *c* In situ hybridization of day 13 mouse fetus with antisense probe to L5 gene. Strong signal is diffusely present in the liver (LV) but no signal is present in the adjacent lung (LG) or heart (HT). (Photomicrograph provided by Drs. Sandra Degen and Jorge Bezerra, Children's Hospital Medical Center, Cincinnati, Ohio.) *d* In situ hybridization for the localization of ADA mRNA in day 16 mouse placenta using ^{35}S labeled antisense RNA probe. Positive signal is indicated by presence of silver grains that appear bright white in dark-field illumination. Strong signal is present in the spongiotrophoblastic layer (ST) and decidua basalis (DB) but not in the chorioallantoic tissue (CA). (*a, b* and *d* are reproduced from the *Journal of Cell Biology* 1991;115:179–190 by copyright permission of the Rockefeller University Press.) *e* in situ hybridization for apoJ mRNA in a day 17 mouse fetus. Intense signal is present in the pancreatic acinar (PA) tissue as well as in the pancreatic duct epithelium (PD). No signal is present in the adjacent small bowel. *f* In situ hybridization for apoJ mRNA in the developing eye of a day 11–12 mouse embryo. The developing lens (LE) and the pigmented epithelial layer (PL) show strong signal and there is weak signal in the primitive cornea (arrowhead). *g* In situ hybridization for apoJ mRNA in the day 17 mouse lung. The signal is distributed peripherally in the lobules where it localizes to the epithelium lining the terminal bronchial derivatives. *h* In situ hybridization for apoJ mRNA in mature postnatal lung tissue. No signal is present in the mature lung. *i* In situ hybridization for apoJ mRNA of day 17 fetal mouse kidney. The epithelial cells lining primitive tubules in the nephrogenic zone show strong signal indicated by black grains in this bright-field illumination (arrowheads). No signal or weak signal is present in the more differentiated tubules or in the developing glomeruli. There is also no signal in the blastemal cells adjacent to the primitive tubules. *j* In situ hybridization for apoJ mRNA in a day 14 mouse fetus. The intense signal is localized to the mucosal epithelium lining the developing ductus cochlearis. *k* In situ hybridization of a transgenic mouse thymus. The section was hybridized with an ^{35}S labeled probe to CAT coding sequences identifying the localization of a transgene construct containing human ADA gene regulatory sequences and a CAT reporter gene. The signal is localized to the thymic cortex where the ADA gene is normally expressed. No signal is present in the medulla (central dark area of the thymus). *l* Phenotypic comparison of day 15 littermates from a mating between mice heterozygous for the mutated c-myb allele. The mutant c-myb mouse is shown on the left, a heterozygous littermate is on the right. The mutant mouse is otherwise normally developed except for the pallid appearance due to severe anemia. (*l* is reproduced from *Cell* 1991;65:677–689, by copyright permission of Cell Press. Photograph provided by Dr. Steven Potter, Children's Hospital Medical Center, Cincinnati, Ohio.)

a, b, i and *l* are bright-field illuminated; all others are dark-field illuminated. Magnification: Panel *a* 25×; *b* 61×; *d* 15×; *c, e, f, j, k* 48×; *g, h* 19×; *i* 95×.

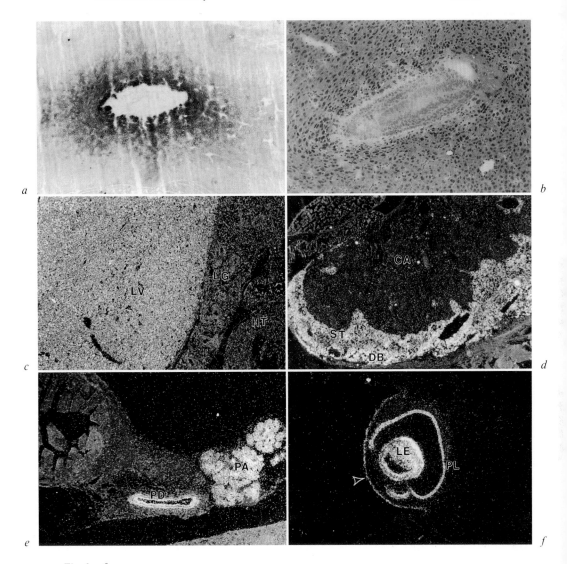

Fig. 1 a–f.

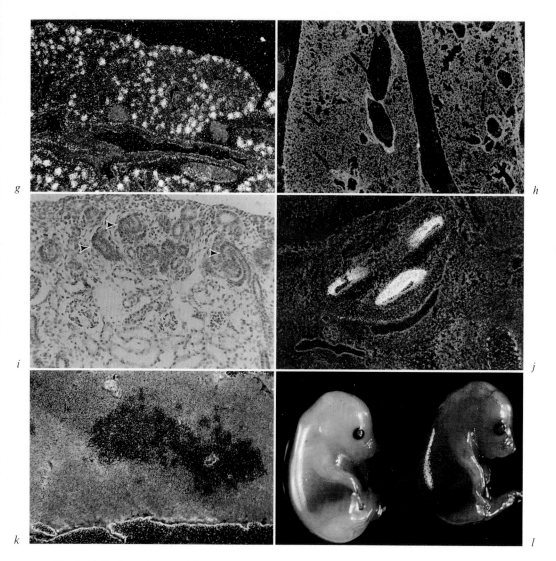

Fig. 1g–l. For legends see p. 10.

Immunohistochemical Localization of Developmental Gene Products

Unfortunately, not all enzymes can be localized histochemically and many important peptides and complex proteins could go unrecognized. Immunohistochemical techniques have enhanced the ability to detect and localize many proteins, including an important class of peptides that has a major impact on developmental processes.

Peptide growth factors were initially identified and characterized by either cell culture systems or in vivo assays, and were identified as small, soluble polypeptides capable of regulating cellular physiology [28–32]. Since Spemann's work [7, 33] led to the concept of an organizer to explain embryonic induction, efforts have failed to identify and isolate any specific chemical agents as the embryonic organizer [34]. The discovery of soluble growth factors and more importantly the identification of growth factor activity in developing fetuses and placentas have identified one class of molecules that may have some of the properties attributed to the putative organizer. In the past decade, many diverse growth factors have been described and implicated in the developmental process [35].

One important growth factor localized during development is transforming growth factor β (TGFβ), originally identified by its ability to stimulate anchorage-independent growth of fibroblastic cells [36, 37]. TGFβ is a member of a large multigene family in species from insects to man [38–43]. TGFβ is an important, multifunctional peptide that controls both growth and differentiation in different cell types and tissues [32, 44]. Several members of the TGFβ family are important in the developmental process. TGFβ affects the developmental process through both growth inhibition in some and stimulation in others [45–47]. Growth factors also influence angiogenesis, chondrogenesis, myogenesis, and interactions between epithelial and mesenchymal cells [47–52].

The localization and distribution of TGFβ in the developing embryo and placenta is restricted to specific tissues but this distribution may change temporally during development. In the human placenta TGFβ has been localized to syncytiotrophoblast and decidua [53]. TGFβ has been identified in the four-cell stage of the preimplantation mouse embryo and the trophectoderm of the blastocyst [54]. In the developing mouse, TGFβ associates closely with mesenchyme or mesenchymal derivatives, particularly during morphogenesis of the palate, larynx, facial mesenchyme, nasal sinuses, meninges and teeth [51]. In the developing mouse heart TGFβ was present in the endocardial cushions but not in the developing myocardium [51]. It has also been identified in epithelial cells of developing bronchioles [55] and in developing hair follicles [51].

Other growth factors have also been implicated in the developmental process by determination of their cellular localization. Basic fibroblast growth factor (bFGF) has been identified in developing heart, striated muscle, and neuroectodermal tissue of rat and chick embryos [56–58]. Platelet-derived growth factor (PDGF), a potent stimulator of fibroblast cell and other cell growth, has been identified in unfertilized eggs and in preimplantation mouse embryos [59, 60]. In the developing mouse, PDGF has been localized to neurons in the spinal cord and dorsal root ganglia [61]. In the developing human embryo, insulin-like growth factors (IGFs) have been localized to hepatocytes, pulmonary epithelium, intestine, kidneys, adrenal cortical cells, muscle fibers, and pancreatic cells [62–64].

The immunohistochemical localization of growth factors in the developing embryo does not establish a causal relationship to the morphogenetic changes occurring during development, but it provides a basis for further study of their role. Based on the localization of TGFβ in decidual and trophoblastic cells, inactivated anti-TGFβ antibodies were used to show a potential importance of TGFβ in controlling trophoblast invasiveness [65]. In a similar study, an antibody to TGFβ inhibited formation of cardiac atrioventricular cushion tissue in vitro [66]. These investigations illustrate the potential for learning about critical developmental events by determining and timing cell-specific activities during embryogenesis.

Cell-Specific Localization of Developmental Genes

Advances in understanding embryogenesis have recently been accomplished by determining specific gene activation during development and the cell specificity of individual activated genes with in situ hybridization (ISH). ISH was initially developed to determine chromosomal localization of different genes [67–69]. The introduction of cDNA probes for use with ISH made available a large array of specific probes to study gene activity [70, 71]. This sensitive, highly reproducible technique has been applied to studying a range of important gene activities that are sometimes unique to the developing embryo, providing information that would otherwise not be obtainable from immunocytochemical analyses alone.

Since many of the growth factors and other proteins identified in developing embryos are secreted, their localization may indicate where it is bound or stored but not necessarily where it is produced. Inappropriate responses in the distant target cells or tissues during development could be a result of defective gene activation or of a defect in the gene itself in cells that secrete the biologically active proteins. Although abnormalities may be apparent in the target tissues, ISH analyses can localize the underlying defect. ISH can also provide a precise time frame of gene activity during development. During embryogenesis, many important gene activities are transient and tightly regulated both temporally and spatially in specific cells. Rapid changes in gene regulation cannot always be detected when the collected data points only identify the final gene product.

ISH also has some technical advantages in the study of developmental processes. Highly specific cDNA or cRNA probes can be generated which, when appropriately hybridized and stringently washed, will remain tightly bound to the target complementary sequences with minimal cross-hybridization to unrelated gene sequences. This contrasts with some immunohistochemical reagents which may show significant nonspecific binding activity or cross-reactivity to similar but unrelated proteins. Some genes or portions of genes have been cloned which have no known function and the sequences do not correspond to any known protein. Nevertheless, using these reagents as probes can identify tissue and cell-specific expression patterns in the developing embryo. This kind of information can identify new genes which may have a critical role in embryogenesis. This kind of approach has been used to study developmental mechanisms using the green alga, *Volvox carteri* [72]. A number of cDNA clones were later identified by Northern blots and ISH as early somatic genes expressed during development. Even when cDNA probes are not readily available, oligonucleotide probes can be synthesized based on partial amino acid sequence data or degenerate probes can be made from sequence data of similar genes in a large gene family or from other species.

The first exploratory use of ISH has been followed by a range of modifications and improvements in the methodology, most significantly in the probes that are now available for application. The various options, briefly covered here, have had a number of excellent reviews [73–75].

Standard nick translation or the random primer method can label double-stranded DNA probes that are complementary to the target mRNA molecule (cDNA probes) [76–78]. A number of available commercial reagent kits can readily generate labeled cDNA probes with high specific activity. These nucleic acid probes can be labeled with radioactive isotopes ^{3}H, ^{35}S, or ^{125}I or with recently developed nonisotopic reagents such as biotinylated nucleotides [79, 80] or digoxigenin-dUTP [81]. Both of these reagents can label DNA probes using nick translation or random primer labeling. A variety of immunohistochemical reagents that specifically recognize either the biotinylated or digoxigenin labeled nucleotides can detect these non-isotopically labeled probes [73].

The cRNA probe has also been extensively used for ISH. This single-stranded RNA transcript 'riboprobe' is complementary to the target DNA or mRNA sequences in the tissue. cRNA probes are synthesized from cDNA templates that are cloned into appropriate plasmid vectors with specific RNA polymerase promotor sites. Using the appropriate RNA polymerase, such as SP6, T7, or T3, a labeled RNA probe is synthesized to be either complementary (antisense strand) or identical (sense strand) to the target mRNA. As with cDNA, radioactive RNA probes can be synthesized and labeled with ^{3}H, ^{35}S, or ^{125}I labeled ribonucleotide or with nonradioactive biotin or digoxigenin. The cRNA probes have several distinct advantages over the more traditional cDNA probes [73, 82]: (1) the single-stranded probe avoids the annealing that occurs between the two strands of a double-stranded probe which results in less efficient hybridization to the target sequences in the tissues; (2) RNA/RNA hybrids are more stable than DNA/RNA or DNA/DNA hybrids; (3) with RNA probes, posthybridization incubation in RNAase removes most unhybridized probe leaving only probe that is hybridized to the mRNA which is protected from digestion, thus greatly reducing background signal. Because of these advantages, riboprobes are 8–10 times more sensitive than standard double-stranded DNA probes [83]. A number of commercial kits are available to synthesize and label cRNA probes as well as providing the appropriate vectors for cloning (Statagene, La Jolla, Calif.; Promega Biotech, Madison, Wisc.; Boehringer Mannheim Biochemicals, Indianapolis, Ind.).

Synthetic oligonucleotide probes, effectively used for ISH [73, 84, 85], are now synthesized by an automated process that is widely available. Oligonucleotides hybridize to their complementary sequence with a high degree of specificity. The synthesized oligonucleotide can be labeled with radioactive isotope, or nonradioactive biotin or digoxigenin conjugated nucleotide by a variety of reactions including 5′ end-labeling, primer extension, and 3′ end-labeling [73, 75]. The advantages of oligonucleotide probes include their ease of synthesis and labeling, high specificity, and small size which may improve tissue penetration [73, 75, 84]. The disadvantages include the lower specific activity of the labeled probes, increased nonspecific hybridization with some probes, and the expense of probe synthesis.

The most recent modification to ISH is in combination with polymerase chain reaction (PCR), thereby adding the dramatically increased sensitivity of DNA PCR amplification (after converting RNA to DNA with reverse transcriptase) [86, 87]. PCR-based ISH will greatly enhance the identification of low copy message beyond the sensitivity of standard ISH.

In situ Hybridization of Embryonic Genes

ISH has been widely used to localize growth factor gene expression during embryogenesis. Expression of IGFs in a variety of embryonic cell types may be regulated by several different mechanisms at the tissue level. IGF mRNAs in human fetal tissues have been localized to perisinusoidal cells in the liver, to perichondrium, and to the connective tissue of several organs [88]. In the developing rat, IGF mRNAs have been localized to the mesenchymal tissue of the head, somites, heart, branchial arches, and limb buds [89]. The regulation of IGF functions are not well understood, but are likely to involve IGF binding proteins that have now been identified and cloned [90]. cDNA probes for ISH have shown that these IGF binding proteins are expressed in the fetus, but the distribution is distinct from that of the IGF genes [91]. These results suggest that the IGFs produced by some embryonic cells may affect distant tissue sites where the local production of IGF binding proteins can modulate the effect.

In the study of the gene expression pattern of TGFβ during development, several TGFβs have been cloned from this large family of related growth factors [92]. The expression patterns for TGFβ_1, β_2, and β_3 have been distinct in the developing mouse [93]. Both the TGFβ_1 and TGFβ_2 genes are expressed simultaneously in the developing heart but with different cellular distributions. The TGFβ_2 gene expresses in the myocardium underlying regions of septation and valve formation but not in atrial or ventricular myocardium. In contrast, the TGFβ_1 gene expresses specifically in the endothelium overlying the presumptive heart valves [93, 94]. The regional localization of both TGFβ_1 and β_2 in the endothelium and myocardium which interact to form cushion tissue may indicate a role for these two closely related growth factor genes in the induction of cardiac development [93, 94].

Potential growth factor genes have been identified and cloned by molecular techniques even before recognizing any function. cDNA probes can identify the tissue which expresses the gene and ISH can determine the cellular localization which in turn may provide some insight into gene function. A previously unidentified gene (L5), recently cloned from a human genomic DNA library, has some homology to the hepatic growth factor gene. Northern analyses localized most L5 expression to the liver with low level expression in lung, adrenal, and placenta [95, 96]. By ISH, L5 gene expression localizes to the hepatocyte [unpubl. data] (fig. 1C), in contrast to hepatic growth factor which localizes to the sinusoidal liver cells [97]. Although the function of this newly identified unique hepatocyte gene in the mature liver has yet to be determined, the highly restricted expression pattern in the fetal mouse liver suggests an importance in liver development.

In addition to growth factors, ISH has localized the expression of a number of other potentially important genes during development. The distribution and localization of growth factor receptors during embryogenesis has identified target tissues of many of the secreted growth factors, e.g., the FGF receptor gene. FGF has been implicated as an important developmental growth factor, with diverse effects including mesoderm induction, induction of homeobox genes, and angiogenesis [98–100]. The recently cloned FGF receptor is widely dispersed during early stages of development in the central neural tissue and retina as well as in a variety of nonneural mesodermal tissues [101]. Although oncogenes were originally recognized for their ability to transform normal cells into neoplastic cells, many oncogenes are homologous to normal cellular genes (proto-oncogenes) that are important in growth regulation and differentiation [102]. Several proto-oncogenes have been localized by ISH in developing embryos with specific temporally and spatially regulated expression patterns [103–106]. Some homeobox genes, with important effects on developmental regulation have been localized by ISH and implicated in the development of the vertebrate eye [107].

The Wilms' tumor gene WT1 is a recessive oncogene that is homozygously deleted in some sporadic and hereditary Wilms' tumors [108–110]. Little is known about the protein encoded by the gene but its predicted amino acid sequence suggests that it may be a transcription factor [109, 110]. mRNA ISH has now shown that, as predicted, WT1 is expressed in the developing kidney [111] and may also have a role in the development of the genital tract, including the uterus, ovary and testis [112]. Recognition of the expanded distribution of WT1 gene expression suggests a broad role in development which may account for the genitourinary tract anomalies occasionally associated with Wilms' tumors [112].

Using ISH to study metabolic proteins that are expressed during implantation and development, we found the entire purine catabolic pathway is coordinately activated in the decidual tissue immediately adjacent to the early implantation embryo (fig. 1A, B). Although previous investigators had shown adenosine deaminase (ADA) activity in placental tissue, localization studies led us to question the role of ADA in the postimplantation reproductive tract. We determined by ISH that ADA gene expression restricts tightly to the decidua basalis and spongiotrophoblastic layer (fig. 1D). In earlier stages it expresses primarily in the decidua capsularis shortly before it regresses [26].

The distribution of the purine catabolic pathway in the decidua suggested a role for this pathway to tissues undergoing programmed cell death as in placental implantation. In seeking other gene expression that may be related to programmed cell death during implantation, we focused on apolipoprotein J (apoJ), a protein in human plasma and other body fluids, with apoJ mRNA in a wide range of tissues [113]. Although the function of apoJ remains controversial, it has been implicated in programmed cell death [114–116]. The distribution and cellular localization of apoJ in the placenta differed from that of ADA and the other purine catabolic enzymes suggesting no relation of its function in the placenta to programmed cell death. ApoJ message localized to the yolk sac epithelium in the placenta but not in decidua. In the fetus, apoJ expressed in several epithelial cell types, e.g. the hepatobiliary system, stomach, pancreas, lung, urinary tract, genital tract, eye, and brain [unpubl. data] (fig. 1E–J). The cellular localization of apoJ expression in the fetus was similar to that in adult organs [unpubl. data] with some notable exceptions. Intense apoJ signal was detected in the fetal lung restricted to the epithelium of the most terminal components of the growing bronchial derivatives as the lung developed (fig. 1G). Postnatal lung had no apoJ message (fig. 1H). In the developing eye, apoJ message was restricted to the lens and the pigmented epithelial layer (fig. 1F) in contrast to the mature eye where message is most prominent in the ciliary body epithelium, the pigmented epithelial layer and in the inner nuclear layers of the retina. In the kidney of the mature mouse, apoJ message appears only in the distal convoluted tubules; in the developing kidney, it appears in the epithelial cells lining the primitive tubules throughout the nephrogenic zone. As the kidney develops and glomeruli and differentiated tubules are formed, the apoJ gene loses expression (fig. 1I).

Although these results do not identify the function of apoJ, the distribution pattern does not suggest a primary function related to programmed cell death. The cellular distribution of apoJ mRNA in adult tissues and other properties of apoJ suggest that apoJ may have a cytoprotective function of particular importance to the epithelial cells of organs exposed to potentially damaging hydrophobic agents in aqueous environments or secretions [113]. Some of the unique distributions of apoJ expression in the developing fetus may reflect special transient conditions or needs of these rapidly growing structures.

Experimental Approaches to Studying Developmental Biology

A large number of specific gene activities arising during development have been identified and characterized by determining cellular localization but without proving or necessarily indicating the role of these genes in development. Based on the localization of specific gene activities during development, well-designed experiments can determine functional roles. Several experimental molecular techniques have been developed and used in this effort.

Taking advantage of oligonucleotides that are antisense to the target mRNA, embryonic tissues incubated in the presence of these reagents will inhibit translation of the mRNA, thereby blocking its effect on the tissue. Using this approach to study the developing heart, Potts et al. [117] showed that blocking TGFβ₃ message inhibited the normal activation of mesenchymal cells during the development of the atrioventricular canal.

The introduction of defined genetic material (transgenes) into the germline of mice allows insertion and expression of genes under controlled conditions to observe the effect of the foreign gene on development, or disrupt normal gene activity in the developing organism [118]. ISH then permits determination of the expression pattern of transgenes linked to a reporter gene such as chloramphenicol acetyltransferase. This approach adapts to studying tissue-specific regulation of genes during development as well as in mature tissue, e.g. the localization of a human transgene that determines expression of the ADA gene in a transgenic mouse thymus (fig. 1K) [119]. In a similar fashion, new techniques have been developed to genetically mark cell clones for introduction into the developing organism for cell lineage studies. Cells genetically labeled with sensitive histochemical markers, such as bacterial β-galactosidase or insect alcohol dehydrogenase, can generate distinct colored reaction products for ready cell identification in a large population of embryonic cells [120].

This application of transgenic technology to the study of developmental biology is illustrated by the c-myb knockout studies. The proto-oncogene c-myb expresses in a variety of malignant tumors derived from neuroectodermal and hematopoietic cell types and in several carcinomas [121–123]. The c-myb gene is also expressed in normal adult and fetal tissue types, including the hematopoietic system [123–125], and in embryonic stem cells suggesting a role in the early stages of development [126]. By homologous recombination in mouse embryonic stem cells, mice homozygous for a mutant c-myb gene were produced [127]. These mice, which appeared normal in the first half of gestation, developed severe anemia by day 15 of gestation. Although the early stages of development were not apparently affected by the mutant c-myb gene, the development of effective erythropoiesis in the fetal liver was severely impaired (fig. 1L). The results suggest that the mutant mice had intact yolk sac stem cells that, following migration to the fetal liver, were incapable of proliferating in the absence of the normal c-myb gene [127].

Conclusions

Understanding the complexity of the developmental process is a formidable task that will require the combined skills and tools of classical embryologists, biochemists, and molecular biologists. Many modern investigative approaches are providing new clues to old problems. Combining standard histological analyses with the examination of individual gene products and gene regulation provides a highly informative picture of the complex processes that determine embryogenesis. This information in turn catalyzes new insights and hypotheses, and suggests strategies to experimentally target specific developmental processes. Clearly, the traditional perspectives of the pathologist can now be expanded to include a detailed cell and tissue analysis of gene expression during embryogenesis. A similar analysis of cell-type specific gene expression will also provide a basis for understanding the course of pathological processes and disease.

Acknowledgments

The authors thank Kathy Saalfeld and Terry Smith for their technical assistance, Judy Huth and Deborah Riddle for assistance in typing the manuscript and Dr. A. James McAdams for his continued support and inspiration. This work was supported in part by a grant from the Society for Pediatric Pathology (D.P.W.) and a Basic Research Grant from the March of Dimes Birth Defects Foundation (B.J.A.).

References

1 Aase JM: Diagnostic Dysmorphology. New York, Plenum Medical Book Co, 1990, pp 5–13.
2 Watson JD, Hopkins NH, Roberts JW, Steitz JA, Weiner AM: The molecular biology of development; in Molecular Biology of the Gene, ed 4. Menlo Park, Benjamin/ Cummings Publishing Co, 1987, vol II, pp 747–750.

3 Berrill NJ: Developmental Biology. New York, McGraw-Hill Book Co, 1971.

4 Van Beneden ME: La maturation de l'oeuf, la fécondation, et les premieres phases du développement embryonnaire des mammiferes d'après des recherches faites chez le lapin. Bull Acad R Belg Cl Sci 1875;40:686–736.

5 Heape W: Preliminary note on the transplantation and growth of mammalian ova within a uterine foster mother. Proc R Soc Lond [Biol] 1890;48:457.

6 Lewis WH, Gregory PW: Cinematographs of living developing rabbit eggs. Science 1929;69:226–229.

7 Spemann H (ed): Embryonic Development and Induction. New Haven, Yale University, 1938.

8 Holtfreter J: Der Einfluss thermischer, mechanischer, und chemischer Eingriffe auf die Induzierfähigkeit von Tritonkeimteilen. Wilhelm Roux Arch Entwicklungsmech Org 1934;132:225–306.

9 Brachet J: Chemical Embryology, ed 2. New York, Interscience, 1950, pp 345–423.

10 Hammond J: Recovery and culture of tubal mouse ova. Nature 1949;163:28–29.

11 Whitten WK: Culture of tubal mouse ova. Nature 1956;177:96.

12 Whitten WK, Biggers JD: Complete development in vitro of the preimplantation stages of the mouse in a simple chemically defined medium. J Reprod Fertil 1968;17:399–401.

13 McLaren A, Biggers JD: Successful development and birth of mice cultivated in vitro as early embryos. Nature 1958;182:877–878.

14 McLaren A: Mammalian Chimeras. Cambridge/UK, Cambridge University Press, 1976.

15 Rossant J, Papaioannou VE: The biology of embryogenesis; in Sherman MI (ed): Concepts in Mammalian Embryogenesis. Cambridge, MIT Press, 1977, pp 1–36.

16 Gardner RL: Mouse chimeras obtained by the injection of cells into the blastocyst. Nature 1968;220:596–597.

17 Jaenisch R, Mintz B: Simian virus 40 DNA sequences in DNA of healthy adult mice derived from preimplantation blastocysts injected with viral DNA. Proc Natl Acad Sci USA 1974;71:1250–1254.

18 Capecchi MR: High efficiency transformation by direct microinjection of DNA into cultured mammalian cells. Cell 1980;22:479–488.

19 Brinster RL, Chen HY, Trumbauer M, Senear AW, Warren R, Palmiter RD: Somatic expression of herpes thymidine kinase in mice following injection of a fusion gene into eggs. Cell 1981;27:223–231.

20 Constantini F, Lacy E: Introduction of a rabbit beta-globin gene into the mouse germ line. Nature 1981;294:92–94.

21 Gordon JW, Ruddle FH: Integration and stable germ line transmission of genes injection into mouse pronuclei. Science 1981;214:1244–1246.

22 Wagner EF, Stewart TA, Mintz B: The human beta-globin gene and a functional thymidine kinase gene in developing mice. Proc Natl Acad Sci USA 1981;78:5016–5020.

23 Hertig AT, Adams EC, McKay DG, Rock J, Mulligan WJ, Menkin JF: A thirteen-day human ovum studied histochemically. Am J Obstet Gynecol 1958;76:1025–1043.

24 Hall K: 5'-Nucleotidase, acid phosphatase and phosphorylase during normal, delayed and induced implantation of blastocysts in mice: a histochemical study. Endocrinology 1971;51:291–301.

25 Knudsen TB, Green JD, Airhart MJ, Higley HR, Chinsky JM, Kellems RD:

Developmental expression of adenosine deaminase in placental tissues of the early postimplantation mouse embryo and uterine stroma. Biol Reprod 1988;39:937–951.

26 Witte DP, Wiginton DA, Hutton JJ, Aronow BJ: Coordinate developmental regulation of purine catabolic enzyme expression in gastrointestinal and postimplantation reproductive tracts. J Cell Biol 1991;115:179–190.

27 Knudsen TB, Gray MK, Church JK, Blackburn MR, Airhart MJ, Kellems RD, Skalko RG: Early postimplantation embryo lethality in mice following in utero inhibition of adenosine deaminase with 2'-deoxycoformycin. Teratology 1989;40:615–625.

28 Ross R, Raines EW, Bowen-Pope DF: The biology of platelet derived growth factor. Cell 1986;46:155–169.

29 Carpenter G, Cohen S: Epidermal growth factor. Annu Rev Biochem 1979;48:193–216.

30 Froesch ER, Schmid C, Schwander J, Zapf J: Actions of insulin-like growth factors. Annu Rev Biochem 1985;47:443–467.

31 Gospodarowicz D, Ferrera N, Schweigerer L, Neufeld G: Structural characterization and biological functions of fibroblast growth factor. Endocr Rev 1987;8:95–113.

32 Sporn MB, Roberts AB, Wakefield LM, Assoian RK: Transforming growth factor-β: biological function and chemical structure. Science 1986;233:532–534.

33 Spemann H, Mangold H: Über Induction von Embryoanlagen durch Implantation artfremder Organisatoren. Wilhelm Roux Arch Entwicklungsmech Org 1924;100:599–638.

34 Hamburger V: The Heritage of Experimental Embryology: Hans Spemann and the Organizer. Oxford, Oxford University Press, 1988.

35 Whitman M, Melton D: Growth factors in early embryogenesis; in Palade G, Alberts B, Spudich J (eds): Annual Review of Cell Biology. Palo Alto, Annual Reviews Inc, 1989, vol 5, pp 93–117.

36 Roberts AB, Anzano MA, Lamb LC, Smith JM, Sporn MB: New class of transforming growth factors potentiated by epidermal growth factor: isolation from non-neoplastic tissues. Proc Natl Acad Sci USA 1981;78:5339–5343.

37 Moses HL, Branum EL, Proper JA, Robinson RA: Transforming growth factor production by chemically transformed cells. Cancer Res 1981;41:2842–2848.

38 Ten Dijke P, Hansen P, Iwata KK, Pieler C, Foulkes JG: Identification of another member of the transforming growth factor type β gene family. Proc Natl Acad Sci USA 1988;85:4715–4719.

39 Jakowlew SB, Dillard PJ, Sporn MB, Roberts AB: Complementary deoxyribonucleic acid cloning of a messenger ribonucleic acid encoding transforming growth factor β from chicken embryo chondrocytes. Mol Endocrinol 1988;2:1186–1195.

40 Kondaiah P, Sands MJ, Smith JM, Fields A, Roberts AB, Sporn MB, Melton DA: Identification of a novel transforming growth factor β (TGF-β5) mRNA in Xenopus laevis. J Biol Chem 1990;265:1089–1093.

41 Padgett RW, St Johnston RD, Gelbart WM: A transcript from a Drosophila pattern gene predicts a protein homologous to the transforming growth factor-β family. Nature 1987;325:81–84.

42 Weeks DL, Melton DA: A maternal mRNA localized to the vegetal hemisphere in Xenopus eggs codes for a growth factor related to TGF-β. Cell 1987;51:861–867.

43 Roberts AB, Sporn MB: The transforming growth factor-betas; in Sporn MB,

Roberts AB (eds): Peptide Growth Factors and Their Receptors: Handbook of Experimental Pathology. Heidelberg, Springer, 1990, vol 95, pp 419–472.

44 Sporn MB, Roberts AB, Wakefield LM, de Crombrugghe B: Some recent advances in the chemistry and biology of transforming growth factor beta. J Cell Biol 1987;105: 1039–1045.

45 Massague J: The TGF-β family of growth and differentiation factors. Cell 1987;49: 437–438.

46 Massague J, Cheifetz S, Ignotz RA, Boyd FT: Multiple type-beta transforming growth factors and their receptors. J Cell Physiol 1987;5(suppl):43–47.

47 Roberts AB, Sporn MB, Assoian RK, Smith JM, Roche NS, Wakefield LM, Heine UI, Liotta LA, Falanga V, Kerhl JH, Fauci AS: Transforming growth factor type-beta: rapid induction of fibrosis and angiogenesis in vivo and stimulation of collagen formation in vitro. Proc Natl Acad Sci USA 1986;83:4167–4171.

48 Rizzino A: Transforming growth factor-beta: Multiple effects on cell differentiation and extracellular matrices (review). Dev Biol 1988;130:411–422.

49 Seyedin SM, Rosen DM, Segarini PR: Modulation of chondroblast phenotype by transforming growth factor-beta. Pathol Immunopathol Res 1988;7:38–42.

50 Florini JR, Magri KA: Effects of growth factors on myogenic differentiation. Am J Physiol 1989;256:C701–C711.

51 Heine UI, Munoz EF, Flanders KC, Ellingsworth LR, Lam H-YP, Thompson NL, Roberts AB, Sporn MB: Role of transforming growth factor-β in the development of the mouse embryo. J Cell Biol 1987;105:2861–2876.

52 Lehnert SA, Akhurst RJ: Embryonic expression pattern of TGF-β type 1 mRNA suggests both paracrine and autocrine mechanisms of action. Development 1988; 104:263–273.

53 Graham CH, Lala PK: Mechanism of control of trophoblast invasion in situ. J Cell Physiol 1991;148:228–234.

54 Slager HG, Lawson KA, van den Eijnden-van Raaij AJM, de Laat SW, Mummery CL: Differential localization of TGF-β₂ in mouse preimplantation and early postimplantation development. Dev Biol 1991;145:205–218.

55 Heine UI, Munoz EF, Flanders KC, Roberts AB, Sporn MB: Colocalization of TGF-beta-1 and collagen I and III, fibronectin and glycosaminoglycans during lung branching morphogenesis. Development 1990;109:29–36.

56 Spirito P, Fu Y-M, Yu Z-X, Epstein SE, Casscells W: Immunohistochemical localization of basic and acidic fibroblast growth factors in the developing rat heart. Circulation 1991;84:322–332.

57 Joseph-Silverstein J, Consigli SA, Lyser KM, Ver Pault C: Basic fibroblast growth factor in the chick embryo: Immunolocalization to striated muscle cells and their precursors. J Cell Biol 1989;108:2459–2466.

58 Gonzales AM, Buscaglia M, Ong M, Baird A: Distribution of basic fibroblast growth factor in 18-day rat fetus: Localization in the basement membranes of diverse tissues. J Cell Biol 1990;110:753–765.

59 Mercola J, Stiles CD: Growth factor superfamilies and mammalian embryogenesis. Development 1988;102:451–460.

60 Rappolee DA, Brenner CA, Schultz R, Mark D, Werb Z: Developmental expression of PDGF, TGF-alpha and TGF-β genes in preimplantation mouse embryos. Science 1988;241:1823–1825.

61 Yeh H-J, Ruit KG, Wang Y-X, Parks WC, Snider WD, Deuel TF: PDGF A-chain

gene is expressed by mammalian neurons during development and in maturity. Cell 1991;64:209–216.

62 D'Ercole AJ, Underwood LE: Growth factors in fetal growth and development; in Novy MJ, Resko JA (eds): Fetal Endocrinology. ORPC Symposia on Reproductive Biology. New York, Academic Press, 1981, vol 1, pp 155–182.

63 Underwood LE, D'Ercole AJ: Insulin and insulin-like growth factors/somatomedins in fetal and neonatal development. Clin Endocrinol Metabol 1984;13:69–89.

64 Han VKM, Hill DJ, Strain AJ, Towle AC, Lauder JM, Underwood LE, D'Ercole AJ: Identification of somatomedin/insulin-like growth factor immunoreactive cells in the human fetus. Pediatr Res 1987;22:245–249.

65 Graham CH, Lala PK: Mechanism of control of trophoblast invasion in situ. J Cell Physiol 1991;148:228–234.

66 Potts JD, Runyan RB: TGFβ blocking antibody can inhibit formation of the AV cushion tissue in vitro. Dev Biol 1989;134:392–401.

67 Gall J, Pardue M: Nucleic acid hybridization in cytological preparations; in Grossman L, Moldave K (eds): Methods in Enzymology. New York, Academic Press, 1971, vol 21, pp 470–480.

68 Gall J, Pardue M: The formation and detection of RNA-DNA hybrid molecules in cytological preparations. Proc Natl Acad Sci USA 1969;63:378.

69 John H, Birnstiel M, Jones K: RNA-DNA hybrids at the cytological level. Nature 1969;223:582–587.

70 Wimber DE, Steffensen DM: Localization of gene function. Annu Rev Genet 1973;7: 205–223.

71 Wiener F, Spira J, Banerjee M, Klein G: A new approach to gene mapping by in situ hybridization on isolated chromosomes. Somat Cell Mol Genet 1985;11:493–498.

72 Tam L-W, Stamer KA, Kirk DL: Early and late gene expression programs in developing somatic cells of *Volvox carteri*. Dev Biol 1991;145:67–76.

73 Lewis ME, Baldino F: Probes for in situ hybridization histochemistry; in Chesselet M (ed): In situ Hybridization Histochemistry. Boca Raton, CRC Press, 1990, pp 1–21.

74 Young WS: In situ hybridization histochemistry; in Bjorklund A, Hokfelt T (eds): Handbook of Chemical Neuroanatomy. New York, Elsevier, 1990, vol 8.

75 Berger SL, Kimmel AR (eds): Methods in Enzymology. San Diego, Academic Press, 1987, vol 152.

76 Rigby PWJ, Dieckmann M, Rhodes C, Berg P: Labeling deoxyribonucleic acid to high specific activity in vitro by nick translation with DNA polymerase I. J Mol Biol 1977;113:237–251.

77 Feinberg AP, Vogelstein B: A technique for radiolabeling DNA restriction endonuclease fragments to high specific activity. Anal Biochem 1983;132:6–13.

78 Feinberg AP, Vogelstein B: Addendum: A technique for radiolabeling DNA restriction endonuclease fragments to high specific activity. Anal Biochem 1984;137:266–267.

79 Langer PR, Waldrop AA, Ward DC: Enzymatic synthesis of biotin-labeled polynucleotides: novel nucleic acid affinity probes. Proc Natl Acad Sci USA 1981;78:6633–6637.

80 Reisfeld A, Rothenberg JM, Bayer EA, Wilchek M: Nonradioactive hybridization probes prepared by the reaction of biotin hydrazide with DNA. Biochem Biophys Res Commun 1987;142:519–526.

81 Baldino F Jr, Lewis ME: Non-radioactive in situ hybridization histochemistry with

digoxigenin-dUTP labeled oligonucleotides; in Conn P (ed): Methods in Neuro-science. New York, Academic Press, 1989, vol 1, pp 282–292.

82 Simmons DM, Arriza JL, Swanson LW: A complete protocol for in situ hybridiza-tion of messenger RNAs in brain and other tissues with radiolabeled single-stranded RNA probes. J Histotechnol 1989;12:169–181.

83 Cox KH, DeLeon DV, Angerer LM, Angerer RC: Detection of mRNAs in sea urchin embryos by in situ hybridization using asymmetric RNA probes. Dev Biol 1984;101:485–502.

84 Montone KT, Budgeon LR, Brigati DJ: Detection of Epstein-Barr virus genomes by in situ DNA hybridization with a terminally biotin-labeled synthetic oligonucleotide probe from the EBV NOT I and PST I tandem repeat regions. Mod Pathol 1990;3: 89–96.

85 Guitteny A-F, Fouque B, Mougin C, Teoule R, Bloch B: Histological detection of messenger RNAs with biotinylated synthetic oligonucleotide probes. J Histochem Cytochem 1988;36:563–571.

86 Nuovo GJ, MacConnell P, Forde A, Delvenne P: Detection of human papilloma-virus DNA in formalin-fixed tissues by in situ hybridization after amplification by polymerase chain reaction. Am J Pathol 1991;139:847–854.

87 Nuovo GJ, Gallery F, MacConnell P, Becker J, Bloch W: An improved technique for the in situ detection of DNA after polymerase chain reaction amplification. Am J Pathol 1991;139:1239–1244.

88 Han VKM, D'Ercole AJ, Lund PK: Cellular localization of somatomedin (insulin-like growth factor) messenger RNA in the human fetus. Science 1987;236:193–197.

89 Stylianopoulou F, Efstratiadis A, Herbert J, Pintar J: Pattern of the insulin-like growth factor II gene expression during rat embryogenesis. Development 1988;103:497–506.

90 Nakatani A, Shimasaki S, Erickson GF, Ling N: Tissue-specific expression of four insulin-like growth factor-binding proteins (1, 2, 3, and 4) in the rat ovary. Endocri-nology 1991;129:1521–1529.

91 Wood TL, Brown AL, Rechler MM, Pintar JE: The expression pattern of an insulin-like growth factor (IGF)-binding protein gene is distinct from IGF-II in the midgesta-tional rat embryo. Mol Endocrinol 1990;4:1257–1263.

92 Melton D: Proceedings of the UCLA Symposium on growth and differentiation factors in development. J Biol Chem, in press.

93 Millan FA, Denhez F, Kondaiah P, Akhurst RJ: Embryonic gene expression patterns of TGF β_1, β_2 and β_3 suggest different developmental functions in vivo. Develop-ment 1991;111:131–144.

94 Akhurst RJ, Lehnert SA, Faissner AJ, Duffie E: TGF-β in murine morphogenetic processes: the early embryo and cardiogenesis. Development 1990;108:645–656.

95 Han S, Stuart LA, Degen SJF: Characterization of the DNF15S2 locus on human chromosome 3: Identification of a gene coding four Kringle domains with homology to hepatocyte growth factor. Biochemistry 1991;30:9768–9780.

96 Degen SJF, Stuart LA, Han S, Jamison CS: Characterization of the mouse cDNA and gene coding for a hepatocyte growth factor-like protein: Expression during develop-ment. Biochemistry 1991;30:9781–9791.

97 Schirmacher P, Geerts A, Pietrangelo A, Dienes HP, Rogler CE: Hepatocyte growth factor/hepatopoietin A is expressed in fat-storing cells from rat liver but not myofibroblast-like cells derived from fat-storing cells. Hepatology 1992;15:5–11.

98 Slack JMW, Darlington BG, Heath JK, Godsave SK: Mesoderm induction in early *Xenopus* embryos by heparin-binding growth factors. Nature 1987;326:197–200.

99 Kemelman D, Kirschner M: Synergistic induction of mesoderm by FGF and TGF-β and the identification of an mRNA codon for FGF in the early *Xenopus* embryo. Cell 1987;51:869–877.

100 Folkman J, Klagsbrun M: Angiogenic factors. Science 1987;235:442–447.

101 Wanaka A, Milbrandt J, Johnson EM Jr: Expression of FGF receptor gene in rat development. Development 1991;111:455–468.

102 Adamson ED: Oncogenes in development. Development 1987;99:449–471.

103 Wilkinson DG, Balles JA, McMahon AP: Expression of the proto-oncogene int-1 is restricted to specific neural cells in the developing mouse embryo. Cell 1987;50:79–88.

104 Cox KH, DeLeon DV, Angerer LM, Angerer RC: Detection of mRNAs in sea urchin embryos by in situ hybridization using asymmetric RNA probes. Dev Biol 1984;101:485–502.

105 Ruppert C, Goldwitz D, Wille W: Proto-oncogene myc is expressed in cerebellar neurons at different developmental stages. EMBO J 1986;5:1897–1901.

106 Wernert N, Raes M-B, Lassalle P, Dehouck M-P, Gosselin B, Vandenbunder B, Stehelin D: c-ets 1 proto-oncogene is a transcription factor expressed in endothelial cells during tumor vascularization and other forms of angiogenesis in humans. Am J Pathol 1992;140:119–127.

107 Monaghan AP, Davidson DR, Sime C, Graham E, Baldock R, Bhattacharya SS, Hill RE: The Msh-like homeobox genes define domains in the developing vertebrate eye. Development 1991;112:1053–1061.

108 Lewis WH, Yeger H, Bonetta L, Chan HSL, Kang J, Junien C, Cowell J, Jones CA, Defoe LA: Homozygous deletion of a DNA marker from chromosome 11p13 in sporadic Wilms' tumor. Genomics 1988;3:25–31.

109 Gessler M, Poustka A, Cavenee W, Neve RL, Orkin SH, Bruns GAP: Homozygous deletion in Wilms' tumors of a zinc-finger gene identified by chromosome jumping. Nature 1990;343:774–778.

110 Call KM, Glaser T, Ito CY, Buckler AJ, Pelletier J, Haber DA, Rose EA, Kral A, Yeger H, Lewis WH, Jones C, Housman DE: Isolation and characterization of a zinc finger polypeptide gene at the human chromosome 11 Wilms' tumor locus. Cell 1990;60:509–520.

111 Pritchard-Jones K, Fleming S, Davidson D, Bickmore W, Porteous D, Gosden C, Bard J, Buckler A, Pelletier J, Housman D, van Heyningen V, Hastie N: The candidate Wilms' tumor gene is involved in genitourinary development. Nature 1990;346:194–197.

112 Pelletier J, Schalling M, Buckler AJ, Rogers A, Haber DA, Housman D: Expression of the Wilms' tumor gene WT1 in the murine urogenital system. Genes Dev 1991;5:1345–1356.

113 Jordan-Stark TC, Witte DP, Aronow BJ, Harmony JA: Apolipoprotein J: A membrane policeman? Curr Opin Lipidol, in press.

114 Montpetit ML, Lawless KR, Tenniswood M: Androgen repressed messages in the rat ventral prostate. Prostate 1986;8:25–36.

115 Buttyan R, Olsson CA, Pintar J, Chang C, Bandyk M, Ng P-Y, Sawczuk IS: Induction of the TRPM-2 gene in cells undergoing programmed death. Mol Cell Biol 1989;9:3473–3481.

116 Sawczuk IS, Hoke G, Olsson CA, Connor J, Buttyan R: Gene expression in response to acute unilateral ureteral obstruction. Kidney Int 1989;35:1315–1319.

117 Potts JD, Dagle JM, Walder JA, Weeks DL, Runyan RB: Epithelial-mesenchymal transformation of embryonic cardiac endothelial cells is inhibited by a modified antisense oligodeoxynucleotide to transforming growth factor β_3. Proc Natl Acad Sci USA 1991;88:1516–1520.

118 Westphal H: Molecular genetics of development studied in the transgenic mouse; in Palade G, Alberts B, Spudich J (eds): Annual Review of Cell Biology. Palo Alto, Annual Reviews Inc, 1989, vol 5, pp 181–196.

119 Aronow B, Lattier D, Silbiger B, Dusing M, Hutton J, Jones G, Stock J, McNeish J, Potter S, Witte D, Wiginton D: Evidence for a complex regulatory array in the first intron of the human adenosine deaminase gene. Genes Dev 1989;3:1384–1400.

120 Lin W-C, Culp LA: Selectable plasmid vectors with alternative and ultrasensitive histochemical marker genes. Biotechniques 1991;11:344–351.

121 Alitalo K, Winquist R, Linn CC, de la Chapelle A, Schwab M, Bishop JM: Aberrant expression of an amplified c-*myb* oncogene in two cell lines from a colon carcinoma. Proc Natl Acad Sci USA 1984;81:4534–4538.

122 Slamon DJ, deKernion JB, Verma IM, Cline MJ: Expression of cellular oncogenes in human malignancies. Science 1984;224:256–262.

123 Torelli G, Venturelli D, Colo A, Zanni C, Selleri L, Moretti L, Calabretta B, Torelli U: Expression of c-*myb* protooncogene and other cell cycle-related genes in normal and neoplastic human colonic mucosa. Cancer Res 1987;47:5266–5269.

124 Westin EH, Gallo RC, Arya SK, Eva A, Souza LM, Baluda MA, Aaronson SA, Wong-Staal F: Differential expression of the *amv* gene in human hematopoietic cells. Proc Natl Acad Sci USA 1982;79:2194–2198.

125 Thiele CJ, Cohen PS, Israel MA: Regulation of c-*myb* expression in human neuroblastoma cells during retinoic acid-induced differentiation. Mol Cell Biol 1988;8: 1677–1683.

126 Dyson PJ, Poirier F, Watson RJ: Expression of c-*myb* in embryonal carcinoma cells and embryonal stem cells. Differentiation 1989;42:24–27.

127 Mucenski ML, McLain K, Kier AB, Swerdlow SH, Schreiner CM, Miller TA, Pietryga DW, Scott WJ, Potter SS: A functional c-*myb* gene is required for normal murine fetal hepatic hematopoiesis. Cell 1991;65:677–689.

Dr. David P. Witte, Department of Pathology,
Children's Hospital Medical Center, Cincinnati, OH 45229-2899 (USA)

Garvin AJ, O'Leary TJ, Bernstein J, Rosenberg HS (eds): Pediatric Molecular Pathology: Quantitation and Applications. Perspect Pediatr Pathol. Basel, Karger, 1992, vol 16, pp 27–98

Peripheral Primitive Neuroectodermal Tumors
Diagnosis, Classification, and Prognosis

Maria Tsokos

Laboratory of Pathology, National Institutes of Health, Bethesda, Md., USA

Introduction

The concept of peripheral neuroectodermal tumors in children and adolescents has recently undergone major changes, and a new terminology has emerged [1]. The term 'primitive neuroectodermal tumor' (PNET) is widely used to describe tumors in both the central and the peripheral nervous systems. To distinguish between the two, Dehner [1] proposed calling one 'central' and the other 'peripheral' PNET. Peripheral PNET in Dehner's classification encompasses conventional neuroblastoma, PNET of bone and soft tissues, and Ewing's sarcoma of bone and soft tissues (fig. 1).

Although classification of conventional neuroblastoma with other peripheral PNET is nosologically correct, sufficient clinical and experimental evidence suggests separating conventional sympathetic neuroblastoma from nonsympathetic peripheral PNET on biologic grounds [2, 3]. In this chapter classic neuroblastoma will be discussed with the other peripheral neuroectodermal tumors, not only because of its topography, but also because comparison with the other peripheral PNET will help explain its unique biology and the need for distinction, in spite of morphologic similarities. The all-encompassing term, neuroblastoma, is not recommended for nonsympathetic peripheral PNET [4].

The peripheral PNET of soft tissue and bone has essentially been rediscovered by the wide use of electron microscopy and immunohistochemistry. After Stout's [5] first decription, peripheral PNET was considered rare, with only 15 cases collected by Nesbitt and Vidone [6] by 1976. Over the

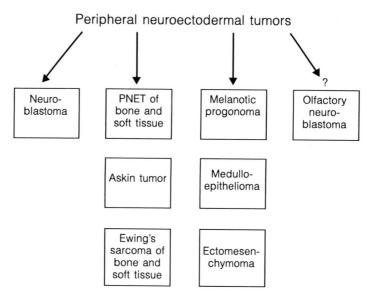

Fig. 1. Classification scheme of peripheral neuroectodermal tumors of bone and soft tissue.

last decade, however, it has been reported in large series [7–14], as isolated cases [15–22], as an entirely osseous variant [23], and now includes tumors of unknown or presumed neural histogenesis, such as the 'small, round cell tumor of the thoracopulmonary region' (Askin tumor) [24–26], and tumors thought to be Ewing's sarcoma with unusual neural features [27, 28].

Ewing's sarcoma is grouped with the neuroectodermal tumors [1, 2] because of similar cytogenetic [29, 30], molecular genetic [31], biochemical [31, 32], and biological [33] features. It has been hypothesized that Ewing's sarcoma and PNET, which express a cholinergic enzymatic profile in vitro [32, 34], arise from postganglionic parasympathetic neurons dispersed throughout the body, while classic neuroblastoma, which is predominantly adrenergic, originates from neurons homing the adrenal gland or the sympathetic nervous system [34]. The main argument against accepting neuroectodermal as the only origin of Ewing's sarcoma is that the available data may come from only a subgroup of tumors already prone to differentiate towards a neural phenotype, and that Ewing's sarcoma may either be heterogeneous, or originate from a primitive cell capable of differentiation along several histogenic lines (neural and mesenchymal). In support of this argument is the existence of a small group of tumors sharing histologic similarities both with Ewing's sarcoma and with matrix-producing, or vasoformative tumors of bone reported with names such as 'polyhistioma' [35], and 'primitive multipotential primary sarcoma of bone' [36], and recently as 'primitive round cell sarcoma of bone' [37].

Three other tumors, pigmented neuroectodermal tumor of infancy (melanotic progonoma), peripheral medulloepithelioma, and ectomesenchymoma, represent distinctive variants of neuroectodermal tumors, that should be kept as a separate subgroup, since they have been much less studied and understood (fig. 1).

The olfactory neuroblastoma (esthesioneuroblastoma) is more closely related to central than to peripheral PNET according to Dehner [1], although it could be either central or peripheral [3], depending on its derivation from the olfactory nerve or the olfactory receptor cells, which are regarded as paraneurons. The origin of olfactory neuroblastoma is generally less well understood than that of conventional neuroblastoma or even peripheral PNET, since this tumor represents neither a simple PNET nor a conventional neuroblastoma [38–43].

In spite of well-defined histologic forms when appropriately differentiated, peripheral PNET may have a primitive histopathologic appearance which requires other than light microscopic methods for diagnosis. The clinical presentation of the peripheral neuroectodermal tumors, with the exception of conventional adrenal neuroblastoma, does not help to distinguish among them. The patients usually present with lytic lesion(s) in bone, with or without extension into adjacent soft tissues, and no other distinctive features. The establishment of therapeutic protocols imposes additional challenges on the pathologist, since accurate histopathologic characterization carries prognostic significance, and is necessary for treatment and future evaluation. The overall survival of patients with Ewing's sarcoma improved with the use of systemic chemotherapy and radiotherapy, even with metastatic disease and truncal primaries [44–47], whereas the survival of patients with advanced stage neuroblastoma is still dismal [48, 49].

In this chapter we will discuss new approaches in the characterization and classification of these tumors, with emphasis on molecular biology, cytogenetics and flow cytometry as they relate to light and electron microscopy and immunocytochemistry.

Light and Electron Microscopy

Neuroblastoma

Assuming neuroblastoma as a dysontogenic neoplasm originating from failure of the neural crest stem cells to differentiate into normal sympathetic neurons clarifies the spectrum of histopathologic differentiation in this tumor. Depending on the degree of cytologic differentiation, three major histologic subtypes are recognized: neuroblastoma, ganglioneuroblastoma, and ganglioneuroma [50].

Neuroblastoma has nests of small round cells with scant cytoplasm and hyperchromatic nuclei, separated by fibrovascular septae (fig. 2a). The tumors are often hemorrhagic, with necrosis or calcification. Rosettes and an intercellular fibrillary eosinophilic material are usually present (fig. 2a, inset). Ganglioneuroblastoma, consisting of primitive and differentiating neuroblasts with or without Schwann cells, is further differentiated into diffuse and composite forms. The diffuse form has a spectrum of neuroblastic differentiation, including mature ganglion cells uniformly distributed throughout the tumor. Ganglionic cell maturation is characterized by cytoplasmic and nuclear enlargement, distinct cytoplasmic borders, increased cytoplasmic eosinophilia, and prominent nucleoli. Nissl substance may be evident as basophilic granules. Intercellular fibrillary material is usually present. The composite form has a background identical to that of ganglioneuroma but harboring discrete areas of primitive neuroblastic cells. Ganglioneuroma consists of terminally differentiated ganglion cells and Schwann cells ensheathing axonal processes which may be myelinated.

The same process of cytodifferentiation in neuroblastic tumors may be the mechanism by which they undergo spontaneous regression. Regression occurs in 8% of all neuroblastomas, 45% of neuroblastomas under 1 year of age, and 87% of stage IV-S neuroblastomas, the last a congenital, disseminated, but mostly favorable form of the disease [51]. Spontaneous differentiation can also occur in metastatic sites (bone, lymph nodes, skin) [52].

Unusual histopathologic forms of neuroblastoma include angiomatoid neuroblastoma with cytoplasmic glycogen [53] and pleomorphic (anaplastic) neuroblastoma [54]. Another form, pigmented neuroblastoma contains neuromelanin, a waste product of catecholamine metabolism [55, 56]. Glycogen, originally regarded as absent, has subsequently been shown in neuroblastoma, although to a lesser extent than in Ewing's sarcoma and PNET [57, 58].

The most consistent and distinctive ultrastructural features of neuro-

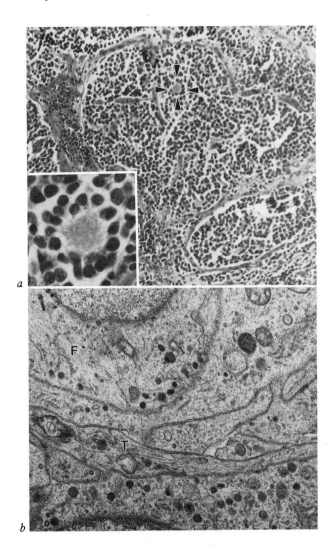

Fig. 2. a In a classic neuroblastoma, aggregates of tumor cells are surrounded by fibrovascular septae. The neurofibrillary material lies in the center of the Homer-Wright rosettes (arrows), shown at higher magnification in the inset. HE. ×103; inset ×260. *b* Overt neuroblastic differentiation by electron microscopy. The tumor cells exhibit cell processes with microtubules (T) and neurosecretory granules in the cytoplasm and the processes. Intermediate filaments (F) in the cytoplasm of the tumor cells are not necessary for diagnosis, but may be present in neuroblastoma and peripheral PNET. Uranyl acetate/lead citrate. ×16,000.

blastoma are cytoplasmic processes with dense core (neurosecretory) granules [50, 58–62] (fig. 2b). The granules are of uniform shape, measure 50–200 nm, and are often in combination with lucent vesicles. Their distribution at the cell periphery and in cell processes helps distinguish them from lysosomes, which are usually close to the Golgi region. A feature of better differentiated neuroblastomas is the circular arrangement of tumor cells with centrally oriented, abortive neural processes that sometimes contain dense core granules, neural tubules and filaments and are the ultrastructural analogue of Homer-Wright rosettes [50]. Other, less specific features include cell molding, which is the result of cell embracement by cytoplasmic processes, cell junctions [59], some of which present as asymmetric synaptic-like junctions, cytoplasmic filaments, and tubules [50, 62].

Ganglionic cells contain a complex arrangement of rough endoplasmic reticulum, many mitochondria, and well-developed Golgi complexes. Various inclusions consist of dense core granules and lysosomes. Myelin figures may be present [63, 64]. Schwann cells have an elongated shape, basal laminae and processes enveloping axons, with or without myelination. Schwann cells identified ultrastructurally in some primitive neuroblastomas may be early indicators of tumor differentiation [61].

Histopathologic Predictive Factors

Traditionally, prognostic factors in neuroblastoma are clinical, such as stage, age, site, vanillylmandelic/homovanillic acid ratio, and increased levels of serum ferritin, and neuron-specific enolase in serum [50]. However, histopathologic criteria alone or in combination with clinical features have also been associated with prognosis [65–69].

A widely used prognosis-related classification scheme has been introduced by Shimada et al. [70]. This scheme emphasizes the participation of the stromal, in addition to the neuroblastic component in the differentiation process, the presence or absence of cytologic maturation in relation to age, and the nuclear mitotic/karyorrhectic index. This index is defined as the percentage of cells with mitotic figures, necrosis and malformed or pyknotic nuclei, when a total of 5,000 cells are evaluated in randomly selected fields. Accordingly, two major histopathologic subtypes of neuroblastoma, the stroma-poor and stroma-rich, are recognized (fig. 3). The word 'stroma' is a misnomer, since it actually refers to the neoplastic Schwann cell component. Both subtypes include cases with favorable and unfavorable outcome. In the stroma-poor subtype, which corresponds to conventional neuroblastoma, the prognosis depends in general on a combination of factors, i.e. age, degree of

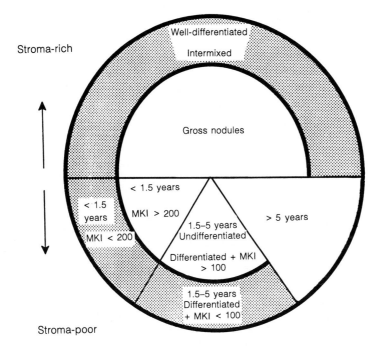

○ Unfavorable prognosis

◉ Favorable prognosis

Fig. 3. In Shimada's classification of neuroblastoma, the two main histologic subtypes depend on the presence or absence of bundles of Schwann cells ('stroma'): stroma-rich = upper half circles; stroma-poor = lower half circles. Tumors of unfavorable prognosis lie inside the white circle. Tumors of favorable prognosis lie in the dotted circle. In the stroma-rich tumors, gross nodules are the only unfavorable prognostic sign. In the stroma-poor group, mitotic/karyorrhectic index (MKI), degree of histologic differentiation and age play important roles in prognosis. After 5 years of age, however, the prognosis is poor for stroma-poor tumors, regardless of maturation and MKI.

maturation, and mitotic/karyorrhectic index. Differentiated histopathology is characterized as 5% or more neuroblastic cells having developed nuclear and cytoplasmic features of ganglion cells, regardless of the presence of fibrillary neuropil.

Specifically, the prognosis in the stroma-poor group is influenced by the mitotic/karyorrhectic index in patients younger than 1.5 years, by the mitotic/karyorrhectic index and degree of maturation in patients between

1.5 and 5 years, and is poor regardless of maturation and mitotic/karyo-rrhectic index in patients over 5 years. In the stroma-rich subtype, which corresponds to the conventional ganglioneuroblastoma/ganglioneuroma, the prognosis is always good, except in those cases with grossly visible discrete nodule(s) of immature cells (nodular). In the favorable stroma-rich tumors, immature neuroblastic cells are distributed uniformly throughout the tumor, either isolated (well-differentiated), or in microscopic clusters (intermixed). The intermixed and nodular subtypes in Shimada's classification correspond to composite ganglioneuroblastoma.

When this scheme is compared to clinical stages, 70% of tumors in the favorable prognosis group correspond to nonadvanced clinical stages (I, II and IV-S) and 80% in the unfavorable group to advanced stages (III and IV). However, patients of the same clinical stage appear to have different outcomes depending on the histopathologic grouping. Gross nodules in stroma-rich neuroblastoma are strongly predictive, even in clinical stages I and II.

This scheme, although cumbersome, was found valid and reproducible in several prospective and retrospective studies [71–73]. When combined with other prognosis-bearing clinical parameters it defines two patient populations: one with a favorable prognosis and survival over 80%, and another with unfavorable prognosis and survival of 20% [74]. Another histopathologic scheme, originated by modification of Shimada's scheme, found that mitotic index is the most important prognostic factor in neuro-blastoma [75].

Ewing's Sarcoma

The light microscopic appearance of Ewing's sarcoma, first described by Ewing [76], only introduces the differential diagnosis of tumors in the Ewing's family, and is by no means sufficient for a final diagnosis. Confirma-tory results are usually obtained by electron microscopic examination, and immunohistochemical staining for myogenous, neural, and occasionally lymphoid markers. The light microscopic criteria of a typical Ewing's sarcoma are summarized in table 1. Typical Ewing's sarcoma consists of sheets of tightly packed, exclusively round, small cells with nuclear homogen-eity, scant cytoplasm, stippled nuclear chromatin and inconspicuous nucleoli, and without any intercellular stroma (fig. 4a). The mitotic rate is very low ($<$ 2 mitoses/10 high power fields).

Table 1. Light microscopic criteria for the diagnosis of typical and atypical Ewing's sarcoma

Features	Typical	Atypical
Architecture	sheets of cells	lobular, alveolar, organoid
Stroma	none	usually present
Cytology	monomorphic	slightly pleomorphic
Cytoplasm	scant	variable
Cell size/shape	small/round/smooth	predominantly small/round (occasional larger or oval cells)
Nucleus	small/round	small/round (larger or oval, or irregular nuclei may be seen)
Chromatin	fine	fine or coarse
Nucleoli	inconspicuous	inconspicuous or prominent
Mitoses	< 2 per HPF	> 2 per HPF

HPF = High power field.

Intracellular glycogen was long considered the light microscopic hallmark of Ewing's sarcoma. However, since glycogen is present in almost all round cell tumors of childhood [57–59, 77–79], and glycogen-negative Ewing's sarcomas constitute one-fifth to one-third of cases [59, 80, 81], its role in the differential diagnosis is limited. On the other hand, large pools of glycogen, characteristically restricted to one or two areas of the cytoplasm, have been found by electron microscopy in almost all Ewing's sarcomas, even in the absence of positive light microscopy [82, 83], and in association with paucity of cytoplasmic organelles, constitute the most characteristic ultrastructural feature of Ewing's sarcoma (fig. 4b).

The ultrastructural appearance of Ewing's sarcoma is unique. The tumor consists of closely apposed homogeneous round cells with smooth contours, diminutive attachments and no extracellular matrix. The tumor cells exhibit very little cytoplasm with sparse organelles, polyribosomes, and pools of glycogen, and small round smooth nuclei with homogeneously distributed heterochromatin [59, 77, 82, 84]. The cytoplasm is generally devoid of cytoskeleton. Principal and secondary tumor cells have been described, corresponding to the light and dark cells by light microscopy [82, 84, 85]. The secondary cells in Ewing's sarcoma have been interpreted as degenerative or viable tumor cells, local reactive stromal cells, or more differentiated forms of principal cells.

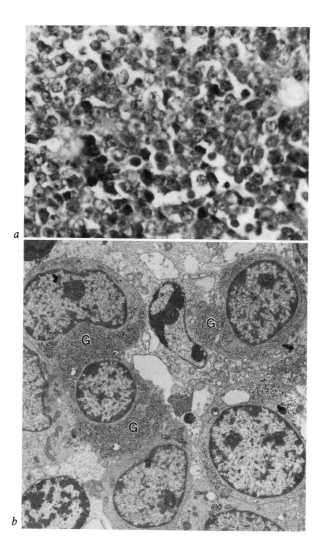

Fig. 4. a Sheets of small, homogeneous, round tumor cells with scant cytoplasm and open chromatin pattern are the characteristic histologic features of a typical Ewing's sarcoma. Nucleoli are discernible, but not prominent. The mitotic rate is very low, or absent. The tumor cells are in close apposition with no intercellular stroma. HE. ×320. *b* Round cells with smooth cytoplasmic and nuclear contours and no intercellular matrix are the ultrastructural features of typical Ewing's sarcoma. The cytoplasm is almost devoid of organelles except for a few mitochondria, and contains large pools of glycogen (G). Intercellular attachments are not evident in this figure. Uranyl acetate/lead citrate. ×4,100.

Variations of the typical Ewing's sarcoma include the large cell (atypical) Ewing's sarcoma, and Ewing's sarcoma with a filigree pattern. The large cell variant contains larger pleomorphic cells with inconspicuous nucleoli and exhibits a high mitotic rate [86]. The filigree pattern consists of delicate interanastomosing bicellular strands of tumor cells in fibrovascular septae [80]. Ultrastructural and immunocytochemical evaluation is important in the diagnosis of the variants, since tumors with a filigree pattern are often diagnosed as something else [87].

Well-formed rosettes [86] and vascular differentiation [81] have been described in the large cell variants of Ewing's sarcoma, in contrast to typical Ewing's sarcoma which may contain pseudorosettes with a collagenous instead of a fibrillary center [80, 81, 84, 88, 89]. Alternatively, the centers of the pseudorosettes may be composed of tumor cell cytoplasms which result from unipolar positioning of cytoplasmic glycogen [89]. Extensive electron microscopic studies have shown that round cell tumors with rosettes or rosette-like formations usually exhibit cell processes with dense core granules [9, 11, 12, 14, 22, 23, 25, 26], and strong neuronal immunocytochemical phenotype [90], and are best classified as PNET.

Atypical Ewing's sarcoma has been defined in several ways, using either light microscopic features alone, e.g. cellular pleomorphism, lobular or alveolar architecture [87], and vascular differentiation [81], or a combination of light and electron microscopic features, e.g. elaborate profiles of cytoplasmic organelles, filaments and desmosome-like structures [85]. Although tumors with unusual histopathologic patterns reminiscent of polyhistioma or primitive multipotential primary sarcoma of bone have been reported as atypical Ewing's sarcoma [50], they are best classified as primitive round cell sarcoma of bone [37]. Similarly, the 'small round blue cell sarcoma of bone mimicking atypical Ewing's sarcoma with neuroectodermal features' [27], and the 'Ewing's sarcoma with neuroblastoma features' [38] are now classified as PNET of bone [23].

We designate as atypical Ewing's sarcoma those tumors that resemble Ewing's sarcoma, but exhibit cellular pleomorphism, high mitotic rate, and a lobular, alveolar, or organoid architecture (fig. 5a). Although some of these tumors may overlap with the large cell (atypical) variant, they lack true rosettes – since this is now considered a feature of PNET – and are not always of a large cell type (table 1). Often, the diagnosis of atypical Ewing's sarcoma cannot be made by light microscopy, but requires electron microscopic and immunohistochemical findings.

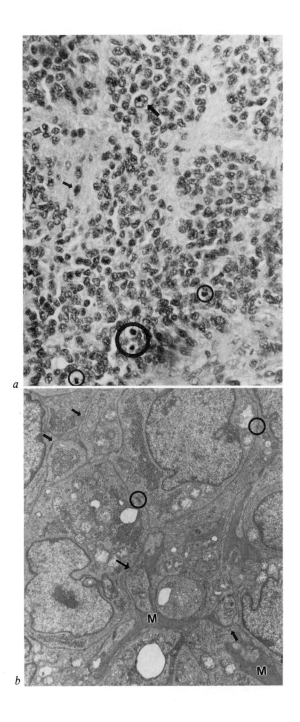

a

b

The ultrastructural features of atypical Ewing's sarcoma are cellular and nuclear pleomorphism, an elaborate content of cytoplasmic organelles (strands of rough endoplasmic reticulum (RER), many mitochondria), a few intermediate cytoplasmic filaments, well-developed attachments, irregular surfaces, and short or long cytoplasmic processes with cell molding, but not dense core granules (fig. 5b). Collagen bundles and a microgranular or amorphous extracellular matrix may be present in the extracellular spaces.

In general, atypical Ewing's sarcoma is neither primitive enough to qualify as a typical Ewing's sarcoma nor does it show overt differentiation into a specific entity. A metastasis from an atypical Ewing's sarcoma differentiated into PNET several years after diagnosis and treatment, while another atypical Ewing's sarcoma exhibited a neural phenotype (processes) in short-term cell culture [unpubl. observation]. These findings and some of the morphologic features of atypical Ewing's sarcoma suggest that it may in reality be a PNET without sufficient light or electron microscopic differentiation. Some atypical Ewing's sarcomas may also show positive staining for neuron-specific enolase, as will be discussed below. The term 'atypical Ewing's sarcoma' is, therefore, used to define tumors with intermediate differentiation between Ewing's sarcoma (very primitive) and PNET (overt ultrastructural focus of neural differentiation), since the degree of neural differentiation may have prognostic implications. These diagnoses have no impact on the current therapeutic protocols, since patients with Ewing's sarcoma, typical or atypical, and peripheral PNET (other than neuroblastoma) receive similar treatment, varying only according to clinical stage, and not according to histology.

Traditionally, extraskeletal Ewing's sarcoma resembles the conventional osseous tumor by light and electron microscopy but often shows a specific architecture, such as rosettes, or organoid, sinusoidal and pericytic patterns [91–98]. Neural features were detected in many tumors diagnosed as extra-

---

Fig. 5. a This atypical Ewing's sarcoma exhibits a nesting pattern. Small aggregates of tumor cells are surrounded by thick collagenous bands. The cytology is slightly pleomorphic (arrows point to larger or oval-shaped cells). This field contains 4 mitotic figures (circles). HE. ×225. *b* The cytoplasmic borders are irregular and form processes which embrace adjacent tumor cells, resulting in cell molding (arrows). Nuclear shape irregularity (infoldings) is present. The number of cytoplasmic organelles is slightly increased. Glycogen may be abundant as in typical Ewing's sarcoma and peripheral PNET. The extracellular electron-dense material (M) is not present in typical Ewing's sarcoma. The short densities at the cellular membranes (circles) represent primitive intercellular attachments. Uranyl acetate/lead citrate. ×3,900.

skeletal Ewing's sarcomas when electron microscopy and immunocytochemistry were employed and suggested that extraskeletal Ewing's sarcoma is a PNET [90, 99–101]. Some tumors with the light microscopic appearance of extraskeletal Ewing's sarcoma showed ultrastructural features compatible with primitive rhabdomyosarcoma [94], and others exhibited myogenous immunohistochemical markers [102], or myogenous differentiation in vitro [103].

We and others [11, 87] believe that the term 'extraskeletal Ewing's sarcoma' has been applied to a heterogeneous group of tumors whose identification relies mostly on electron microscopic and immunocytochemical examination. Hartman et al.'s [87] reclassification of Ewing's sarcoma of distal extremity as typical Ewing's sarcoma, atypical Ewing's sarcoma, peripheral PNET, rhabdomyosarcoma, synovial sarcoma, and primary sarcoma of bone, emphasizes the need for strict light and electron microscopic, and immunocytochemical criteria for Ewing's sarcoma. Tumors with rosettes, processes with neurosecretory granules, or neural markers by immunocytochemistry [90] should be classified as PNET, and tumors with cells exhibiting ample pink cytoplasm, a fair amount of cytoplasmic filaments with or without Z-band material and myogenous immunohistochemical markers [77] should be classified as rhabdomyosarcoma, reserving sheet-like, or large lobular areas and homogeneous, round, primitive cytology to extraskeletal Ewing's sarcoma. Terms such as 'primitive soft tissue sarcoma' [94] or 'sarcoma of undetermined histogenesis' [104] have been proposed for those sarcomas which are more differentiated than Ewing's sarcoma, but not sufficiently to bear a specific diagnosis.

Having excluded PNET, neuroblastoma and rhabdomyosarcoma, one is, in practice, left with Ewing's sarcoma (typical or atypical), some primitive sarcomas of soft tissue and bone, and malignant lymphoma. Certain types of lymphoma, especially in extranodal sites, may be difficult to distinguish from Ewing's sarcoma by light microscopy. Ultrastructurally however, conventional lymphoma lacks intercellular attachments, and the nuclei contain marginated chromatin [105]. Most lymphomas do not contain large amounts of glycogen. Scanning electron microscopy shows cells with ruffled surfaces in lymphoma contrasting with the smooth cytoplasmic surfaces in Ewing's sarcoma [106].

Small round cell sarcomas of bone and soft tissue, with areas simulating Ewing's sarcoma, but also with neoplastic cartilage, bone, vascular structures, epithelium, and hemangiopericytoma-like areas have been reported under the terms 'polyhistioma' [35], and 'primitive multipotential primary sarcoma of bone' [36]. The clinical characteristics, i.e. patient's age, racial incidence, and propensity for soft tissue extension, were similar to those of Ewing's sarcoma. Related tumors, focally exhibiting a Ewing's sarcoma-like cytology plus well-differentiated islands of cartilage or osteoid and cartilage, have been reported under the terms 'mesenchymal chondrosarcoma' [107] and 'small cell osteosarcoma' [108].

A few tumors of bone, and soft tissue, referred to the National Cancer Institute with a diagnosis of Ewing's sarcoma, contained areas of extracellular dense collagenous or myxoid matrix (fig. 6a, b) composed of collagen bundles and proteoglycan particles by electron microscopy (fig. 6c), but without the overt osteoid, or cartilaginous differentiation to justify the diagnosis of mesenchymal chondrosarcoma, or small cell osteosarcoma. Most had a lobular architecture or a filigree pattern. Two tumors that lacked extracellular matrix contained portions with a sinusoidal or a pseudohemangiopericytomatous pattern in which the vessels did not have open lumina and lacked the so-called staghorn appearance requiring identification with a reticulin stain. Focal hermangiopericytomatous differentiation, i.e. cells with long processes and reduplicated basal lamina, was noted ultrastructurally in one of them, and abortive lumina formation in the other. The tumor cells lacked Weibel-Palade bodies.

The architectural differences were accompanied by cytologic differences. Larger, more pleomorphic cells with abundant pink or clear cytoplasm and a coarse nuclear chromatin pattern were the main cell type, with some spindle cells. The mitotic rate was variable, but always higher than that of typical Ewing's sarcoma. Tumor cells exhibited a spectrum of differentiation by electron microscopy. Some tumor cells were very primitive with sparse organelles (Ewing's sarcoma-like), and others contained a fair amount of cytoplasmic organelles and prominent nuclear pleomorphism. Variable amounts of glycogen, prominent intracytoplasmic intermediate filaments, and conspicuous intercellular attachments were present. Cell molding and processes were not prominent. Neurosecretory granules were never observed. Elongated tumor cells exhibited abundant RER and peripherally condensed filaments, resembling myofibroblasts.

We have grouped these tumors under the generic term, 'primitive round cell sarcoma of bone' [37], since we have not encountered elements other than mesenchyme to qualify them as polyhistioma, or multipotential sarcoma, although we believe them to be similar. These tumors constitute only a very small proportion of round cell tumors of bone and soft tissue, but their distinct light and electron microscopic features warrant their distinction

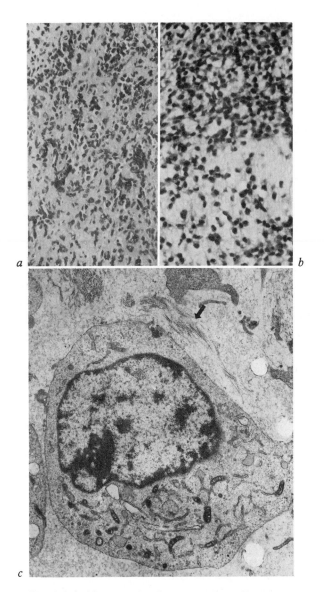

Fig. 6. Primitive round cell sarcoma. Both illustrated tumors arose in bone and contained areas composed of solid sheets of small round cells with minor pleomorphism. *a* This tumor exhibited areas with a thick intercellular collagenous matrix, in which tumor cells invaded haphazardly, without any specific arrangement. HE. ×105. *b* This tumor showed foci in which tumor cells were loosely arranged in a myxoid intercellular matrix. A pink cytoplasm and an eccentrically placed nucleus was present in some tumor cells. HE. ×190. *c* Representative electron micrograph from the tumor shown in figure 6b. The

from typical and atypical Ewing's sarcoma. Such tumors may prove to be the link between Ewing's sarcoma and mesenchymal tumors, such as chondrosarcoma, small cell osteosarcoma, angiosarcoma, or hemangiopericytoma. Interestingly, Ewing's sarcoma has osteogenic ability when implanted into the musculature of nude mice [109].

The only such tumor found in soft tissue had a focal sinusoidal pattern and mild vasoformative characteristics by electron microscopy, simulating Bednar's solid dendritic cell angiosarcoma, a variant of extraskeletal Ewing's sarcoma [92], but without the same degree of vascular differentiation. Similar cases with abortive Weibel-Palade bodies have been reported as Ewing's sarcoma of bone with endothelial character [110], reprising Ewing's original concept of an endothelial origin (diffuse endothelioma of bone) [76].

Histopathologic Predictive Factors

There are no well-established histopathologic criteria with prognostic significance in Ewing's sarcoma, as there are in neuroblastoma. Clinical characteristics such as anatomic location, serum lactic acid dehydrogenase (LDH) levels, and extent of disease at diagnosis relate individually to clinical outcome [111, 112], but not after adjustment for differences in the composition of treatment groups [112]. Soft tissue extension is associated with metastatic disease at presentation and poor survival [113]. Tumor volume at presentation and responsiveness to chemotherapy have been the two major prognostic factors in patients with nonmetastatic osseous Ewing's sarcoma [45].

Among histopathologic features, a filigree pattern and necrosis were associated with a dismal prognosis [80, 81, 114, 115]. Patients with conventional Ewing's sarcoma of extremities fared better (p = 0.03) than those with peripheral PNET, rhabdomyosarcoma, and synoviosarcoma, although the numbers of patients were small [87].

Primitive Neuroectodermal Tumor

The common term 'peripheral PNET' has prevailed over those of peripheral neuroepithelioma or adult neuroblastoma and is currently used

---

extracellular matrix is composed of small particles (proteoglycans) and small clusters of thin collagen bundles (arrow). The tumor cell lays loose in this matrix, and contains well-developed RER and mitochondria. In some tumor cells the RER was remarkably dilated. Uranyl acetate/lead citrate. ×5,850.

almost exclusively for PNET of bone and/or soft tissue, including the neuroectodermal tumor of bone, and the malignant round cell tumor of thoracopulmonary region (Askin tumor).

The light microscopic appearance of peripheral PNET may closely resemble an atypical Ewing's sarcoma. Although true rosettes have been considered mandatory for the diagnosis of a PNET [8, 9, 23, 116, 117], they are usually focal [23, 25] or absent [12, and personal experience], even in PNET with ultrastructural neural features. Whenever present, the rosettes are of the Homer-Wright or the Flexner-Wintersteiner type [8, 11, 13, 89]. Most tumors exhibit a lobular, organoid, or pseudoalveolar pattern or sheets of round cells as in Ewing's sarcoma [12–14, 26]. Schmidt et al. [13] distinguished patterns resembling typical and atypical Ewing's sarcoma, neuroblastoma, rhabdomyosarcoma, and hemangiopericytoma. Some PNET, including the Askin tumor, have an organoid pattern with pseudo-rosette formation [26, 89] (fig. 7a, b). Pseudorosettes lack the central neurofibrillary material of true rosettes, having instead a center composed either of collagenous material or of the cytoplasmic portions of the converging tumor cells.

The predominant cell type of peripheral PNET is round, like that of Ewing's sarcoma, but with more cytoplasm, round nucleus with fine or coarse chromatin and inconspicuous or prominent nucleoli. In a few tumors, oval and spindle cells may be focally present. Mitotic figures are moderate to numerous [8, 12].

Despite histogenetic and ultrastructural similarities with neuroblastoma, peripheral PNET does not usually show mature elements, such as ganglion cells, nerve bundles, and mats of neuropil [89, 118]. Neuropil, whenever present, usually appears only in the center of the Homer-Wright rosettes, and not free in the intercellular space [9–11, 22]. Those few tumors with ganglion-like cells in primary [11, 12, 90, 119] or recurrent sites [15] showed a more advanced stage of differentiation with intervening neurofibrillary stroma as in classic neuroblastoma, and should accordingly be classified as neuroblastomas. In examples, the ganglioneuroblastic elements were juxtaposed to primitive small round cell areas [11], mingled with the primitive neuroepithelial cells forming the Homer-Wright rosettes, or clustered in the stroma [119]. Astrocytic cell differentiation in one of these tumors made it strikingly similar to a cerebellar medulloblastoma with neuronal and neuroglial differentiation. One tumor with ganglionic cell differentiation [116] also contained ependymal cells, cartilage, and smooth muscle, and has been retrospectively interpreted as a malignant ectomesen-

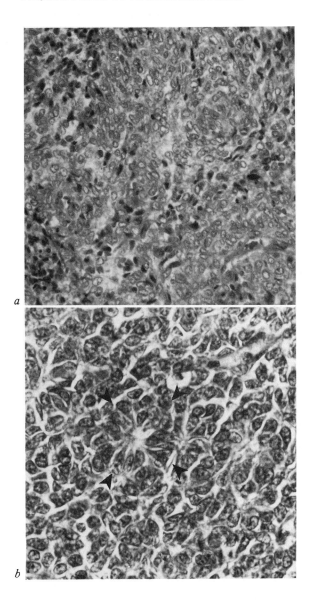

Fig. 7. Peripheral PNET have different patterns. *a* Clusters of cohesive, predominantly oval-shaped cells form vague whorls. HE. ×175. *b* A more organoid pattern exhibits pseudorosettes (arrowheads). In contrast to the rosettes in figure 2, the center of the pseudorosettes is composed of the cytoplasm of the converging cells, and lacks neurofibrillary material. HE. ×340.

chymoma [120]. In the interpretation of ganglion cell-containing peripheral PNET, their limited occurrence, particularly with other cellular components, requires consideration of other diagnoses, such as conventional neuroblastoma, medulloepithelioma, or ectomesenchymoma.

Epithelial cell differentiation has been reported only in three PNET, two with clusters of epithelial-like cells [8, 121], and one with well-formed nonciliated glandular structures with mucin [122], which also exhibited astrocytic cells, cells with a rhabdoid phenotype, and occasional cells with cilia-like structures. The epithelial structures in a PNET are puzzling. Among neuroectodermal tumors, epithelial cell differentiation occurs in peripheral nerve sheath tumors (schwannomas) [123–125]. Although PNET is considered to be a neuroepithelial tumor, well-formed glandular structures are rare. One explanation for this phenomenon would be origin from a multipotential, nonneural crest committed, germinal neuroepithelial cell. Although alternative hypotheses, such as origin from an uncommitted mesenchymal cell [126], cannot be entirely excluded, the electron microscopic, immunocytochemical and cytogenetic data support a neuroectodermal cell rather than a mesenchymal cell origin.

We conclude that the diagnosis of peripheral PNET is not always easy to make, especially by light microscopy. The majority of PNETs have a light microscopic appearance similar to that of atypical and sometimes typical Ewing's sarcoma and the only diagnostic light microscopic criterion, the presence of true rosettes, is only rarely encountered. Therefore, the ultimate distinction between atypical and typical Ewing's sarcoma and PNET lies predominantly in their ultrastructural appearances. Positive staining for a neural marker is suggestive of PNET, but ultrastructural confirmation is required, because of lack of specificity of most neural markers. Thanks to the wide use of electron microscopy, the diagnostic accuracy has improved greatly since Willis [127] considered as unproven all Ewing's sarcoma cases unless a complete autopsy excluded the possibility of a metastatic neuroblastoma. We now classify most Ewing's sarcomas as peripheral PNETs and no longer require association of a PNET with a peripheral nerve, as originally suggested [6, 128].

The ultrastructural hallmark of PNET is cytoplasmic processes containing neurosecretory (dense core) granules [9, 15, 23, 25, 26, 50]. Since the granules are fewer and more variable in shape and size than in classic neuroblastoma (fig. 8a, b), their proper identification as neurosecretory is more difficult than in conventional neuroblastoma. The issue becomes more complicated since neurosecretory granule-like structures have been described even in lymphomas

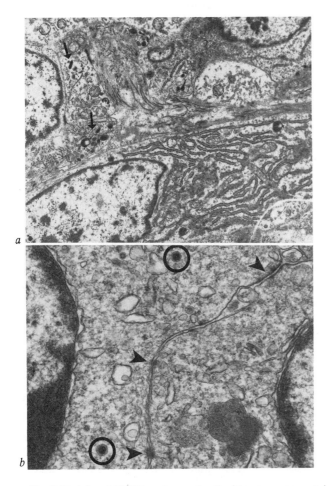

Fig. 8. Peripheral PNET. *a* A round cell with processes containing neurosecretory granules (arrows) is present on the left, and an elongated cell with a fair amount of RER occupies the bottom part of the picture to the right. Intercellular matrix with collagen bundles is present in the extracellular space. Uranyl acetate/lead citrate. ×7,100. *b* Neurosecretory granules (circles), and short intercellular attachments (arrowheads). The granules exhibit a regular shape and size and a peripheral halo, but are fewer than those in neuroblastoma. Uranyl acetate/lead citrate. ×22,000.

[129]. In general, the overall neural differentiation in PNET is much less evident than in neuroblastoma. Most often, the tumor is predominantly composed of cells without processes. The cells contain cytoplasmic glycogen, like those in Ewing's sarcoma, although with a more elaborate content of

cytoplasmic organelles (RER, mitochondria) and possibly intermediate cyto-plasmic filaments. Neural ultrastructural features have been consistently reported in many ultrastructural studies of PNET, although with variable degrees of differentiation and dense core granules ranging from rare and atypical to abundant and typical [7, 9–12, 14, 15, 22, 23, 25, 26, 50].

Identification of cells with neural characteristics is important to make the distinction from typical and atypical Ewing's sarcoma. Even the focal presence of cells with neural processes containing unequivocal neurosecre-tory granules is sufficient for this distinction. Although cell molding and processes suggest a PNET, only the characteristic neurosecretory granules substantiate the diagnosis. Otherwise, the tumor is called atypical Ewing's sarcoma. Failure to identify neurosecretory granules in some tumors with processes is probably due to lack of tumor differentiation rather than sampling error, because similar results are obtained in both needle core and open biopsies. Needle core biopsies have been sufficient to distinguish between typical Ewing's sarcoma and PNET or atypical Ewing's sarcoma. We believe that Ewing's sarcoma expresses a real spectrum of ultrastructural differentiation which is not secondary to sampling problems. Our current terminology (typical and atypical Ewing's sarcoma and PNET) aims at preserving this histologic differentiation for future analysis and clinicopatho-logic correlations.

Other important, but not primary diagnostic features are filaments and microtubules [10, 15, 23, 25]. Basal lamina-like material and long-spaced collagen (Luse bodies) lie in the intercellular spaces [9, 10]. We have also encountered extracellular pools of electron-dense and microgranular materi-al, as well as long-spaced collagen in some PNETs, in contrast to Ewing's sarcoma, which is devoid of extracellular matrix. Pools of an extracellular and endoplasmic microgranular material in a chest wall tumor may have identified it in retrospect as a PNET (i.e. Askin tumor), although it lacked overt ultrastructural evidence of neural differentiation [130]. Variably deve-loped intercellular attachments are present [12, 15, 25], although their development is usually superior to that in typical Ewing's sarcoma. Termin-ally differentiated ganglion or Schwann cells have not been seen, although elongated cells with a fair amount of RER suggest early Schwannian differen-tiation (fig. 8b).

Distinction of PNET from rhabdomyosarcoma usually relies on cell processes with neurosecretory granules in the former and cytoplasmic fila-ments with Z-band material in the latter. Immunohistochemistry is of further substantial support in such a distinction. The distinction of PNET from neuroblastoma is theoretically impossible on ultrastructural grounds alone, although most neuroblastomas express a more differentiated neural pheno-type than PNETs. Immunohistochemistry and molecular genetic markers offer further assistance for this distinction.

Medulloepithelioma

Medulloepithelioma is a unique, rare, peripheral PNET exhibiting im-mature neural canals [131]. The tumor exhibits potential for differentiation along neuronal, glial, and ependymal lines. In contrast to peripheral PNET of soft tissue and bone, it resides mainly in the intracranial cavity, and in intraocular sites, where it presumably arises from the medullary epithelium of the optic vesicles [125, 132]. Chondroblastic and rhabdomyosarcomatous elements have been observed in the ocular medulloepithelioma [132]. Me-dulloepithelioma has also been described in the sacrum [133], probably originating from persistent medulloepithelial vestiges, and in the gonads, as part of primitive gonadal tumors [134].

Melanotic Neuroectodermal Tumor of Infancy

The melanotic neuroectodermal tumor of infancy, otherwise known as melanotic progonoma or retinal anlage tumor, is an unusual tumor of the head and neck, occurring mainly in the maxilla (70% of cases), but also in the mandible, skull, femur, epididymis, skin, brain, uterus, ovary and mediastinum [135–138]. Of 178 cases reported through 1985 [139], only 9 had a malignant course, 2 of which had arisen in the genital tract of adult females.

The tumor may exhibit a nesting or alveolar pattern in a prominent fibrocollagenous stroma or grow in sheets of cells. The solid pattern of growth may be associated with an infiltrating or metastasizing component of the tumor [138]. The cell types are a small hyperchromatic, neuroblastic cell, and a larger pigmented cell with eosinophilic or amphophilic cyto-plasm and a vesicular nucleus [135–142]. Schwann cells and nerves, or neuronal cell differentiation have been observed [138, 140].

By electron microscopy, the melanotic neuroectodermal tumor of infancy exhibits diverse and more differentiated cell types than does the conventional peripheral PNET plus a different age distribution. The melanotic neuroectodermal tumor usually occurs in children less than 1 year old [135, 138], and the PNET in adolescents and adults [8, 12, 14, 26], although melanotic progonoma has occurred in adults [136, 143]. The histogenetic origin of this tumor remains controversial, although biochemical, ultrastructural, and histochemical data support a neural crest origin [135, 136, 138, 139].

A mixture of cell types is observed ultrastructurally, the most prominent of which is the pigmented cell, which contains abnormal melanosomes in various stages of development. Clusters of pigmented cells are surrounded by basal lamina [135, 138, 140, 142, 144]. Other cell types are small neuroblastic cells with processes, fibroblastic cells [138, 140], and intermediate cells containing melanosomes, neurosecretory granules and microtubules. Intermediate cells also exhibit incomplete basal lamina on the outer surface [141].

Olfactory Neuroblastoma

With the exception of a few examples arising in the nasopharynx, maxillary sinuses, or cranial cavity [39, 145], olfactory neuroblastoma is a tumor of the nasal cavity, originating superiorly in the region of the cribriform plate or laterally above the middle turbinate. Although first called esthesioneuroepithelioma [146], it has been known as esthesioneuroblastoma and olfactory neuroblastoma [38, 39, 78, 145]. A bimodal age distribution has peaks in age groups 11–20 years and 51–60 years [147].

In spite of its designation, olfactory neuroblastoma does not always exhibit the light microscopic appearance of conventional neuroblastoma. Taxy et al. [148] recognized two groups: group I resembled conventional neuroblastoma composed of sheets and nests of small round cells in a fibrillary stroma with Homer-Wright rosettes and dystrophic calcification; group II exhibited neuroepithelial features and was composed of confluent sheets or anastomosing cords of intermediate-sized cells with foci of necrosis and desmoplasia. Fibrillary background was absent, but occasional true Flexner rosettes were present. Epithelial differentiation, including squamous cell differentiation, is occasionally observed [149]. Other histopathologic features are a biphasic epithelial and stromal pattern, and formation of papillae.

Ganglion cell differentiation is rare [148]. Schwann cell differentiation is not readily apparent by light microscopy, despite fusiform cells reminiscent of Schwann cells, especially at the tumor/stroma interface [78, 148]. Homer-Wright rosettes, although helpful in diagnosis, are not always present [148, 150]. Pigment, present in a few cases [150, 151], may increase considerably after heterotransplantation into nude mice [41].

Histopathologic characteristics including tissue necrosis have no prognostic significance [148, 150–153], and efforts to subclassify olfactory neuroblastoma according to prognosis have been largely abandoned, although the general degree of differentiation and the number of mitoses appeared to be main prognostic factors in one study [154]. Although complete surgical excision has been considered the most important determinant of prognosis [150], Kadish et al. [155] reported that the tumor is radiosensitive and potentially radiocurable when limited to the nasal cavity and paranasal sinuses (clinical groups A and B). A favorable prognosis in early stage disease (groups A and B) with a single modality of treatment (surgery or radiation) has been reported by Elkon et al. [147]. Stage C patients, on the other hand, in general had a poor prognosis. Local recurrence is the usual result of treatment failure [147, 155].

The most consistent ultrastructural features are processes and dense core granules [38, 42, 43, 145, 148]. Schwann cells are also frequently present, despite their inconspicuousness by light microscopy [148, 156]. Rosettes and pseudorosettes may occasionally be seen [145]. The tumors may contain numerous dystrophic axons [157], or acinar spaces, formed by cells with microvilli and attenuated apical cytoplasm with slender processes and bulbous expansions as in the olfactory vesicle [18]. Melanosomes are present in the pigmented form of olfactory neuroblastoma [41].

Ectomesenchymoma

Ectomesenchymoma is a rare malignant tumor of soft tissues, or central nervous system [158–165], thought to arise from remnants of migratory neural crest cells (ectomesenchyme). The predominant peripheral sites are the head and neck region of infants [158, 161, 162], although the tumor may occur in other sites, and has been reported in 4 adults from Thailand [122, 159, 160, 163–165]. The tumor is traditionally composed of well or poorly differentiated neuroblastic cells (neuroblastoma, ganglioneuroblastoma, ganglioneuroma, PNET) and one or more malignant mesenchymal elements, usually rhabdomyosarcoma [159, 161]. Other reported elements are benign and malignant neuroglial and Schwann cells, liposarcoma, melanocytes, malignant fibrous histiocytoma, meningioma, and chondrosarcoma [160]. With the frequent use of electron microscopy and immunohistochemistry, primitive neuroectodermal elements are now discovered in tumors that would otherwise be classified as rhabdomyosarcomas, and malignant rhabdomyoblasts are readily seen in predominantly neuroblastic tumors [unpubl. data]. It is therefore conceivable that the frequency of malignant mesenchymoma will increase in the future and that mixed neural and mesenchymal phenotypes will not be as rare as previously thought. Although malignant ectomesenchymoma may be considered a subtype of malignant mesenchymoma, it is unique in its containment of neuroectodermal elements and it stands as a distinct clinicopathologic entity [158, 161].

Ultrastructural studies of ectomesenchymoma have shown neuroblastic and rhabdomyoblastic differentiation [160] plus other elongated, noncommitted cells with RER [unpubl. observations]. The tumor cells are not closely apposed and are separated by extracellular matrix composed mainly of collagen bundles.

Immunohistochemistry

General Considerations

Immunohistochemistry is not always conclusive for a neuroectodermal tumor, since the available antibodies against neural antigens either lack specificity or fail to stain the most primitive tumors in the group. In contrast, the immunohistochemical markers of rhabdomyosarcoma are both specific and sensitive. As a general rule, therefore, a tumor positive for both neural and muscle markers and having compatible histopathology is best regarded as rhabdomyosarcoma.

Immunohistochemical results should always be interpreted in association with light and electron microscopy. Regarding the contribution of the three morphologic methods, light microscopy suggests, immunohistochemistry supports, and electron microscopy confirms the diagnosis of a round cell tumor of childhood.

Immunohistochemistry has mainly a diagnostic, but also a prognostic [3, 166] role. The prognostic value stems from the detection of antigens normally associated with increasing differentiation and of oncogene-associated proteins. Histogenetic associations of immunohistochemical markers [90, 167–170] are not wise however, since several traditional assumptions have changed, e.g., cytokeratin expression is not limited to epithelial cells, its being expressed also by rhabdomyosarcoma [89, 171–173], and leiomyosarcoma [174].

The most important antigenic determinants in the round cell tumors of childhood and their value in the differential diagnosis will be discussed.

Neural Markers

Neural markers in the diagnosis of round cell tumors can be characterized as neuronal and neuroendocrine (table 2). The neuronal markers are neuron-specific enolase (NSE), neurofilament triplet proteins (NFTP) with molecular weights of 200, 160, and 68 kilodaltons (kd), two of the microtubule-associated proteins (MAP-1 and MAP-2) and the β-tubulin. The neuroendocrine markers are chromogranin, synaptophysin, and vasoactive intestinal peptide (VIP).

Neuron-Specific Enolase

NSE is a glycolytic enzyme occurring in several homodimeric or heterodimeric isoenzymes, formed by subunits α, β and γ [175]. The γγ-isoenzyme, originally identified as a brain-specific protein (14-3-2), was considered unique to neuronal structures and was named 'neuron-specific enolase' (NSE) [176]. NSE was considered to be specific for certain neuroendocrine cells and tumors [177, 178], and helped identify a neural histogenesis of several neoplasms, such as the so-called small cell tumor of thoracopulmonary region [26], and the neuroectodermal tumor of bone [23]. Recently, however, a sensitive enzyme immunoassay has shown that γ-enolase, although highly enriched in brain, is also widely distributed in tissues, especially in its αγ form [175]. Similar results have been obtained by immunohistochemical methods, which have revealed γ-enolase-containing cells in several tissues [179] and tumors [180], challenging the neuronal specificity of NSE [175, 181].

Schmechel [175] has suggested combining immunohistochemical NSE localization with biochemical measurement of tumor tissue extracts, and Seshi and Bell [182] have introduced monoclonal antibodies, one of which is specific for neural tissues, and does not react with nonneuronal enolase in solid phase radioimmunoassay. However, biochemical studies may be complicated by the inclusion in specimens of normal structures rich in NSE, such as nerves, resulting in false-positive findings, and monoclonal may be less

Table 2. Neural and neuroendocrine markers in peripheral neural tumors compared with other round cell tumors of childhood

Tumors	NSE	NFTPs			MAP1/MAP2	Synapto-physin	Chromo-granin
		200	160	68			
NB	+	+[1]	+	–[2]	+	+	+
ES	–	–	+	–	–	–	–
AE	+[3]	–	+	–	ND	ND	ND
PRCS	+[3]	–	–	–	ND	ND	ND
PNET	+	–	+	–	ND	–	–
Olf. NB	+	+[4]	+[4]	+[4]	ND	ND	ND
Melanotic progonoma	+	ND	ND	ND	ND	ND	ND
Ectomesenchymoma	+	–	+	–	ND	ND	ND
RMS	+	+	+	+	–[5]	–[6]	–[6]
Lymphoma	–	–	–	–	–	–	–

NB = Neuroblastoma; NSE = neuron-specific enolase; ES = Ewing's sarcoma; NFTP = neurofilament triplet proteins; AE = atypical Ewing's; PRCS = primitive round cell sarcoma; MAP = microtubule-associated proteins; PNET = primitive neuroectodermal tumor; Olf. NB = olfactory neuroblastoma; ND = not done; RMS = rhabdomyosarcoma.
[1] More differentiated tumors (ganglioneuroblastoma/ganglioneuroma) positive.
[2] One commercial antibody investigated by us did not stain paraffin sections. On frozen sections however, NFTP-68 is expressed by neuroectodermal primitive tumors more often than the other two NFTPs.
[3] Some cases, but not all, stain positive.
[4] Some tumors positive for neurofilaments. The specific subtypes (200, 160 or 68) were not investigated.
[5] Only well-differentiated rhabdomyoblasts stain positive.
[6] In some studies (see text) some rhabdomyosarcomas were positive for synaptophysin and chromogranin.

sensitive than polyclonal antibodies, despite their increased specificity [183]. Available polyclonal antibodies against NSE do not stain all tumors. Lymphoma and well-defined conventional Ewing's sarcoma of bone are negative for NSE [178]. Although skeletal and extraskeletal Ewing's sarcoma are occasionally positive for NSE [11, 59, 90, 99, 101, 114, 184], this may be the result of neuroectodermal differentiation. In our series, typical Ewing's sarcomas by light and electron microscopy have been negative for NSE, while several atypical Ewing's sarcomas have been positive (table 2).

Artifactual, diffuse, light staining of B5-fixed tissues for NSE [178] may explain positive staining of lymphomas, which are often fixed in B5. Decalcification interferes with demonstration of NSE [178].

Since NSE-positive cells have been observed in primitive round cell sarcomas of bone [37], and mesenchymal chondrosarcoma [185], a matrix producing tumor of bone should not be excluded on the basis of NSE staining alone. Since both primitive and differentiated cells of some rhabdomyosarcomas may stain positively [59, 178], staining for muscle markers is needed to differentiate PNET from rhabdomyosarcoma.

The advantage of NSE as a neural/neuroendocrine marker lies in its consistent staining of central and peripheral neural tumors, regardless of the degree of histologic differentiation [178, 186–188]. Consequently, even less well differentiated tumors, such as PNET of soft tissues, bone and chest wall stain positively for NSE [9–14, 16, 23, 25, 26, 178]. The few NSE-negative tumors even in the presence of neural differentiation by ultrastructure [11] suggest failure due to artifactual changes.

Neurofilament Triplet Proteins

Neurofilaments form part of the neuronal cytoskeleton, and are present exclusively in neuroectodermal cells [189]. Preparations of isolated mammalian neurofilaments contain three proteins with molecular weights of approximately 200, 160, and 68 kd, named neurofilament triplet proteins (NFTP) [190]. Monoclonal antibodies specific for the three subunits are commercially available and have been used in the diagnosis of neural and neuroendocrine neoplasms [186, 191–194].

All three NFTPs are positive in neuroblastoma, ganglioneuroblastoma, ganglioneuroma [186, 192], and olfactory neuroblastoma [148, 194]. In general, NFTP have been more frequently demonstrated in ganglioneuroblastoma and ganglioneuroma than in neuroblastoma in paraffin-embedded material [186, 193]. More differentiated cells stain selectively with antibodies against the 200- and 160-kd proteins [186, 192, 193], similar to the enhanced expression of neurofilament RNA during differentiation of neuroblastoma cells in vitro [195]. Thus the value of high and intermediate weight neurofilaments in the diagnosis of poorly differentiated neuroectodermal tumors is limited. In contrast, NFTP-68 has been demonstrated even in primitive neuroblastic cells, and in normal neural embryogenesis, in which low molecular weight precedes high molecular weight NFTP expression [196]. However, staining with several anti-NFTP-68 antibodies has been variable [10, 12, 90].

The method of fixation influences NFTP staining with adverse effects from prolonged formaldehyde fixation (more than 24 h) [197]. Although all neuroblastomas stained positive for NFTP in frozen sections, variable results were obtained with formalin-, or alcohol-fixed paraffin-embedded tissues [186, 187, 192]. Osborn et al. [192] reported that only axons in ganglioneuroblastomas were positive in alcohol-fixed paraffin-embedded tissues. It has been speculated that tumors with low levels of NFTP expression, i.e., poorly differentiated neuroblastomas, are more likely to be positive in frozen than in paraffin sections.

We used commercially available anti-NFTP antibodies to stain formalin-fixed, paraffin-embedded neuroblastoma, peripheral PNET and Ewing's sarcoma [198]. All tumors were consistently negative for NFTP-200, except for selective axonal and ganglionic cell staining in a few neuroblastomas. Antibodies against NFTP-160 weakly stained all neuroblastomas, PNET and a few Ewing's sarcomas irrespective of degree of differentiation. This weak, sometimes equivocal staining even in Ewing's sarcomas has suggested to us a nonspecific staining. The anti-NFTP-68 antibody failed to stain any of the tumors or the nerve control sections, although axonal processes in control sections or entrapped in tumor tissue were intensely positive for the other two NFTP. Trypsinization improved mildly the staining with anti-NFTP-200, had a negative effect on the staining of anti-NFTP-160, and did not influence the staining of anti-NFTP-68, which remained negative in all cases. These commercially available NFTP antibodies have very little value in the diagnosis of PNET, neuroblastoma and Ewing's sarcoma when formalin-fixed, paraffin-embedded tissues are used [9, 10, 199]. Another antibody directed against high and medium molecular weight NFTP has become available for which preliminary data show a greater sensitivity than the others in paraffin-embedded tissues.

Normal skeletal and smooth muscle stain with antibodies against all three NFTP proteins, including NFTP-68. The 68-kd protein is a component of nonneuronal cells and of skeletal myofibrils [200]. Rhabdomyosarcoma, Ewing's sarcoma, and malignant fibrous histiocytoma have been reported to be positive for NFTP-68 [172, 201, 202].

Nuclear staining with anti-NFTP-200 was consistent, enhanced with trypsin, and observed in all types of tumors, irrespective of cytoplasmic staining. Although otherwise unexplained, nuclear staining for NFTP may result from cross-reactivity with nuclear lamins, and should not be interpreted as positive staining of the tumor for NFTP. The lamins are proteins localized to the inner surface of the nuclear envelope and show structural homology to cytoskeletal proteins [203]. Cross-reactivity suggests that lamins share epitopes with NFTP, since antigenic determinants of intermediate filaments may be both unique and shared [204].

β-Tubulin and Microtubule-Associated Proteins

Microtubules mediate a broad range of biologic functions. Neuronal microtubules contribute to elongation and stabilization of neuritic processes by polymerization of tubulin subunits [205]. α- and β-tubulins are the two major soluble microtubule proteins, each occurring in about six isotypes. The class III β-tubulin isotype in birds is restricted to neurons of central and peripheral nervous systems [206]. A monoclonal antibody (TU27) recognizes an epitope common to all β-tubulin isotypes, and another one (TUJ1) a unique, neuron-associated epitope of class III β-tubulin [207]. TUJ1 has identified primitive neuroepithelial cells in an ovarian teratoma and may be a diagnostic marker of primitive neuroectodermal cells [207].

Microtubules possess accessory proteins, the so-called microtubule-associated proteins (MAP), which were originally defined by their ability to promote the in vitro assembly of tubulin [208]. MAP-1, MAP-2 and τ have been isolated from neural tissues and have also been shown in cultured neuroblastoma cell lines [209]. Although antibodies against MAP-1 and MAP-2 occur widely in nonneuronal tissues, the primary structure of the molecule is neural-tissue specific [210]. MAP expression in childhood tumors was positive in neuroblastoma and negative in Ewing's sarcoma, malignant lymphoma, and rhabdomyosarcoma, although well-differentiated rhabdomyoblasts were positive, like mature striated muscle cells [199]. Because MAP-1 and MAP-2 were present even in poorly differentiated neuroblasts and were far more sensitive than NFTP, and even NSE [199], they may prove useful as differential markers.

Neuroendocrine Markers

Synaptophysin

Synaptophysin, a glycoprotein localized to the membrane of presynaptic vesicles [211], is localized in neurones, paraganglia and several epithelial neuroendocrine cells, as well as in neuroblastomas, paragangliomas, carcinoids and neuroendocrine carcinomas [187, 191, 212, 213]. Immunohistochemical studies have shown decreased synaptophysin positivity in paraffin-embedded as compared to frozen tissues [187]. The most appropriate fixative for synaptophysin may be 95% alcohol [214], although conventional fixatives, including buffered formalin, may be suitable provided that exposure to the fixative is no more than 2–4 h.

Almost all neuroblastomas, including olfactory neuroblastoma, contain cells positive for synaptophysin using commercially available antibodies [187, 191, 212, 213]. The positive cells are well-differentiated neuroblasts and ganglion cells [187, 191], while most poorly differentiated cells are negative [187], resembling antigen expression during normal embryogenesis and differentiation [215]. A few examples of Ewing's sarcoma, rhabdomyosarcoma and lymphoma were negative for synaptophysin [212, 213]. Using paraffin-embedded tissue and commercially available antibodies, synaptophysin was demonstrable by us and by others in occasional cells in a few PNETs [10]. In contrast to other studies, we have consistently found a higher expression of synaptophysin in rhabdomyosarcoma than in PNET. The results are unexplained, unless some rhabdomyosarcomas have a partially neuronal phenotype as illustrated by their expression of NSE and NFTP.

Chromogranin

Chromogranins, a group of acidic proteins, have been divided into three classes – A, B, and C [216]. All three chromogranins are present in human chromaffin granules, in which they form a major part of the soluble matrix proteins. Immunoelectron microscopy has identified chromogranins in neurosecretory granules of neuroendocrine tissues [217], confirming their role as universal markers of neuroendocrine tissues and their tumors [217, 218]. Tumors positive for chromogranin contain neurosecretory granules, e.g., neuroblastoma, pheochromocytoma, pulmonary small cell carcinoma, and Merkel cell carcinoma of the skin [187, 191, 212].

All neuroblastomas have been positive for chromogranin in frozen sections [191], but the immunoreactivity was predominantly localized in ganglion cells. Despite variable results with formalin-fixed, paraffin-embedded tissues, more differentiated cells were generally positive [187, 212]. Numbers of cytoplasmic neurosecretory granules and staining for chromogranin has a positive correlation [219]. We find chromogranin of no diagnostic value in less well differentiated neural tumors, such as most cases of PNET and its variants, although others found that 50% of PNET were positive [10]. This discrepancy may be due to differentiation, since Homer-Wright rosettes were considered mandatory for diagnosis in these series and all cases exhibited abundant neurosecretory granules. Other tumors, e.g. lymphoma and Ewing's sarcoma, were negative for chromogranin, and rhabdomyosarcoma had occasional, weakly stained cells [212].

Vasoactive Intestinal Polypeptide (VIP)

VIP, normally present in neural tissues, is localized in neurons and nerve fibers by immunofluorescence [220]. Its biologic and functional significance is unknown, although it may act as a neurotransmitter [221], and it has induced morphologic differentiation of neuroblastoma cells in vitro [222]. Ganglioneuroblastoma and ganglioneuroma secrete VIP in the watery diarrhea syndrome [223]. An immunohistochemical study of neuroblastoma, ganglioneuroblastoma, ganglioneuroma, and pheochromocytoma showed VIP in ganglion cells but not in primitive or poorly differentiated neuroblastic cells and pheochromocytes, suggesting an association of VIP with ganglionic cell differentiation and maturation [224].

Supportive Cell Markers

S-100 Protein

S-100 protein, originally considered a marker of normal glial and Schwann cells, of tumors derived from these cells, and of malignant melanoma [225–227], has been identified in other benign and neoplastic tissues, such as various carcinomas and sarcomas [185, 228, 229], including rhabdomyosarcoma [59, 171], and leiomyosarcoma [229]. Positive cells in rhabdomyosarcoma are usually large differentiated cells [171]. S-100 protein is expressed in variable percentages in almost all round cell tumors of childhood, except for classic Ewing's sarcoma of bone, and nonhistiocytic lymphoma (table 3).

S-100 protein stains well-differentiated Schwann cells in ganglioneuroblastomas and ganglioneuromas [185, 226, 227]. Although poorly differentiated neuroblastoma was originally thought to be negative [225–227], S-100-positive, spindle-shaped cells were subsequently found in the thin collagenous septa, which separated the neuroblastic nests, and were associated positively with prognosis [166]. Olfactory neuroblastoma also exhibits S-100-positive cells [39, 41, 148], which form a marginal network similar to classic neuroblastoma, and correspond to cells with Schwannian ultrastructural features at the tumor stroma interface [39, 148, 188]. In a pigmented olfactory neuroblastoma, the S-100 protein-positive cells were distributed throughout the tumor, which when heterotransplanted into nude mice showed positivity for S-100 protein in newly appearing large epithelioid and in isolated small pigmented cells [41].

Peripheral PNET variably exhibits S-100 protein-positive cells in a

Table 3. Supportive cell and intermediate filament markers in the common peripheral neural and other small round cell tumors of childhood

Tumors	S-100	GFAP	Keratins	Vimentin	Desmin	Muscle actin
NB	$+^1$	$+^2$	–	$–^3$	–	–
ES	–	–	+	+	–	–
AE	–	–	ND	+	–	–
PRCS	$+^4$	ND	ND	+	–	–
PNET	$+^4$	$+^2$	ND	+	–	–
Olf. NB	$+^1$	+	+	ND	–	–
Melanotic progonoma	–	ND	+	+	+	+
Ectomesen-chymoma	$+^4$	$+^5$	$–^6$	+	+	+
Rhabdo-myosarcoma	+	–	+	+	+	+
Lymphoma[7]	–	–	–	+	–	–

Key: same as table 2.
[1] In primitive tumors positive cells are localized in fibrous network. In differentiated tumors Schwann cells positive.
[2] A few exceptional cases positive (see text).
[3] Only a few exceptional cases.
[4] Patchy pattern throughout the tumor.
[5] Only glial cells positive.
[6] Rhabdomyoblasts stained positive in one case.
[7] Nonhistiocytic.

patchy pattern distributed throughout the tumor [9, 12–14, 25], expressed in as many as 50% of the tumors [8–10, 12, 14, 23, 95]. These results may be explained on the basis of variable, but usually poor, Schwann cell differentiation in a given PNET. The S-100 protein-positive cells are usually very few even in the positive cases, and exhibit either a spindle, or a round cell phenotype [25, and personal experience]. Although S-100 protein-positive cells suggest differentiation to precursor Schwann cells, overt Schwannian differentiation, as in neuroblastoma, is absent. A similar pattern of S-100 protein staining has been observed in the so-called extraskeletal Ewing's sarcoma [95, 101], and the primitive round cell sarcoma of bone [37]. S-100 protein-positive cells in primitive round cell sarcomas of bone may connote early cartilaginous differentiation. Although S-100 protein staining of overtly differentiated cartilaginous tumors has been documented [230], the data concerning staining of primitive round cells in mesenchymal chondrosarcomas are conflicting and limited to positivity in rare primitive cells in one tumor [231], and to well-differentiated lacunar and transitional cells in another [185].

The melanocytic cells in the pigmented melanocytic tumor of infancy were found surprisingly negative for S-100 protein in two studies [139, 141].

Glial Fibrillary Acidic Protein (GFAP)

GFAP is an intermediate filament protein considered to be specific for astrocytes, ependymal cells [232, 233], and for tumors originating from them [234]. GFAP should be absent, therefore, from peripheral neural tumors, including the peripheral PNET [9, 10, 23, 186, 212]. The immunoreactive glial filaments occasionally detected in Schwann cells and nonglial tumors [235, 236] has been attributed to nonspecific cross-reacting antibodies to shared determinants among the various intermediate filament classes [204].

A few GFAP-positive peripheral PNET cannot be explained on the basis of cross-immunoreactivity of antibodies (table 3). GFAP-positive tumor cells in two NFTP-negative poorly differentiated neuroblastomas remained positive even after preabsorption of the GFAP antisera with a great excess of NFTP supporting tumor multipotentiality [186]. Molenaar et al. [191] detected GFAP but not NFTP in fiber bundles and not in primitive cells by double immunostaining in 3 cases of ganglioneuroblastoma/ganglioneuroma, illustrating GFAP staining of differentiated Schwann cells. Recent studies have identified GFAP-positive nonmyelinating Schwann cells, and GFAP-negative myelinating Schwann cells respectively [237].

GFAP-positive cells in an olfactory neuroblastoma resembled astrocytes, and were seen in addition to the S-100 protein-positive spindle cells [188], as in some normal Schwann cells in the olfactory nerves, which are positive for GFAP and resemble astrocytes [238]. Other olfactory neuroblastomas have been negative for GFAP [239].

Among the peripheral PNET, the rare GFAP-positive cells showed overt astrocytic cell differentiation, but were in tumors with ganglion cells [119], or glandular differentiation [122], which may not fit in the conventional category of PNET. A GFAP-positive PNET [12], also positive for vimentin and neurofilaments, may have been the result of cross-reactivity of antibodies. GFAP-positive cells were detected in an ectomesenchymoma which also showed gliomatous differentiation [120]. There are no reports of GFAP-positive typical or atypical Ewing's sarcoma, rhabdomyosarcoma, or lymphoma.

Other Markers for Intermediate Filaments

Keratin/Desmoplakin/Desmoglein

Keratins/cytokeratins (CKs) comprise a large family of closely related polypeptides encoded by multiple genes [240]. Although originally thought to characterize tumors of epithelial origin, CKs occur also in nonepithelial cells [89, 171–174, 241]. CK and NFTP coexpress in certain neuroendocrine tumors [242]. Among the tumors discussed here (table 3), olfactory neuro-blastoma may be positive [38, 148] or negative [41] for either or both markers. Olfactory neuroblastoma has foci of CK-positive cells in both its neuroblastic (group I), and neuroendocrine (group II) histologic subtypes [148]. Conventional neuroblastoma and ganglioneuroma are negative for CK [243, 244]. Peripheral PNFT however, may contain CK-positive cells [unpubl. data]. The immunohistochemical characterization of the pigmented neuroectodermal tumor of infancy is limited, but in one study, the epithe-lioid melanocytic cells were positive and the small neuroblastic cells negative for CK [141].

CK8- and CK18-positive cells have been detected in Ewing's sarcoma by immunofluorescence of frozen tissues [201] as well as in a Ewing's sarcoma cell line [245], and confirmed by two-dimensional gel electrophoresis, West-ern blotting of cytoskeletal proteins and electron and immunoelectron microscopy [201]. With rare exception [114], CK-positive cells in Ewing's sarcoma had escaped detection in formalin-fixed tissues [169, 246, 247] suggesting either masking of the epitopes by aldehyde fixation, or limited sampling. Possibly influencing the appearance of CK-positive cells in Ew-ing's sarcoma is the degree of histologic differentiation. New appearance and increased levels of CK were observed in Ewing's sarcoma cell lines treated with dibutyryl cyclic AMP, retinoic acid and phorbol ester [245], and in a tumor that had been negative for CK prior to heterotransplantation into nude mice [247].

Rhabdomyosarcoma also expresses CK8 and CK18 by immunohisto-chemistry [171, 172] and Western blot analysis [173]. Although CK-positive cells in sarcomas are unexplained, rare CK-synthesizing cells spontaneously appear in nonepithelial cell cultures, suggesting spontaneous losses of genes or deregulation of CK gene expression [248].

Desmoplakins, specific major plaque proteins that connote desmosomal differentiation [249], have been reported in a punctate pattern in Ewing's sarcoma, especially in association with vimentin [201]. This rare phenome-non may be used as an additional diagnostic marker of Ewing's sarcoma.

Desmoglein, another desmosomal associated protein [250], was not present in 3 Ewing's sarcoma cell lines even after differentiation into a neural phenotype or after treatment with phorbol ester which decreased the neural markers and increased CK expression [245].

Vimentin

Vimentin, originally considered a marker for mesenchymal cells, expresses widely among eukaryotic cells [251], so that it cannot discriminate among the tumors in the differential diagnosis of PNET [95, 169, 201, 244, 246, 252, 254, 255] (table 3), but it can serve to evaluate preservation of tissue antigenicity [243]. However, variability in preservation of vimentin antigenicity with various fixatives (better with alcohol), and variable staining with several antivimentin antibodies (polyclonal versus monoclonal, different commercial source) may affect this evaluation [244].

Ewing's sarcoma (osseous and extraosseous), rhabdomyosarcoma, and primitive round cell sarcoma of bone generally express abundant vimentin filaments with diffuse cytoplasmic staining [37, 95, 114, 169, 201, 212, 243, 252]. Vimentin staining in PNET is variable, present in 50% of the cases [9, 12]. In pigmented neuroectodermal tumor of infancy, the neuroblastic cells were negative and the melanocytic cells positive for vimentin [141].

Heterogeneous staining in rhabdomyosarcomas reflects normal muscle differentiation in which vimentin is predominant in early stages, later replaced by desmin [253]. The majority of undifferentiated small cells are positive for vimentin, while differentiated cells are negative for vimentin and positive for desmin [254].

Immunostaining of lymphomas for vimentin has varied from 0 to 70% [243, 244, 255] and neuroblastoma is usually negative, with a few exceptions [212, 243, 244]. The presence of vimentin in PNET and its absence from neuroblastoma, especially the ganglion cells, may relate to maturation, similar to the maturation-associated switch from vimentin to neurofilament synthesis in normal neural development [256]. Although cultured neuroblastoma cells have been found positive for vimentin, vimentin positivity of various cell types is a generalized phenomenon in vitro, and is not specific for neuroblastoma [257].

Desmin and Muscle Actin

Desmin- and muscle-specific actin, cytoskeletal proteins with immense diagnostic value, are found almost exclusively in myogenous tissues and tumors [252, 254, 258–261]. However, anomalous expression of desmin in nonmyogenous cells, such as stellate cells of rat liver [262] and rat astrocytes of Müller glia [263], has been reported. Desmin-positive cells have been rarely detected in central [264] and peripheral [102] PNET for unexplained reasons. In general, different antidesmin antibodies give different results in paraffin sections, the polyclonal being more sensitive than the monoclonal antibodies. Prolonged formaldehyde fixation results in negative desmin staining [197].

The antibody against muscle-specific actin is reported to be more sensitive than desmin in the diagnosis of rhabdomyosarcoma [212, 258], but a commercially available antimuscle-specific actin antibody has been found in nonmyogenous tumors, such as malignant fibrous histiocytoma, and schwannoma [265].

In spite of the difficulties, commercially available antibodies against desmin- and muscle-specific actin are useful and widely used in the identification of even primitive rhabdomyosarcomas, which are very often positive, in contrast to other tumors which are almost always negative [11, 101, 114, 169, 185, 212, 252, 259–261]. A few desmin- and muscle-specific actin-negative rhabdomyosarcomas may result from either suboptimal fixation, or very primitive differentiation [171, 254]. Both desmin- and muscle-specific actin are also expressed by smooth muscle cells, myofibroblastic cells and their tumors [258, 261]. Antibodies against a sarcomeric actin however, stain exclusively skeletal muscle and rhabdomyosarcoma, for which they are both specific and sensitive [266]. Use of both antiactin and antidesmin antibodies is recommended in conjunction with other muscle markers (myoglobin, creatine phosphokinase MM, skeletal muscle myosin) for optimal results [260, 267].

Cell Surface Antigens

Of the several monoclonal antibodies against known and unknown membrane determinants from neuroblastoma, lymphoma/leukemia, and Ewing's sarcoma, most are not tissue-specific, requiring combinations of appropriate panels of antibodies in a diagnostic setting (table 4).

Hematopoietic Cell Antigens
The leukocyte common antigen (LCA), a major constituent of the membrane of hematopoietic cells, establishes the lymphoreticular/hematopoietic origin of a round cell tumor with 93–94% sensitivity in paraffin-embedded

Table 4. Cell surface antigens in the differential diagnosis of peripheral neuroectodermal tumors from other small round cell childhood tumors

Tumors	HSAN 1.2	W6/32	HNK-1	HBA-71	UJ127.11	UJ13A/ UJ167.11	KP-NAC8	LCA	BA-1	BA-2	MB2	Mab5A7	FMG25
NB	+	$-^1$	+	-	+	+	+	-	+	+	$-^2$	+	+
ES	-	+	+	+	-	-	+	-	+/-	+/-	+	-	-
PNET	+/-	+	+	+	+	+	ND	-	+/-	-	+	+	ND^3
Rhabdo	-	+	+	$-(+)^4$	-	+	-	-	+/-	+	-	-	-
Lymphoma/ leukemia	-	+	-	$-(+)$	$+^5$	-	$+^5$	+	+	+	$+^6$	-	$+^7$

[1] Ganglion cells and stage IV-S NB are positive (see text).
[2] Only ganglion cells positive.
[3] Two out of 2 medulloblastomas were positive.
[4] Only occasional cases positive, mainly in rhabdomyoblasts.
[5] Positive leukemia cell lines.
[6] B-cell lymphomas.
[7] Mainly T-cell lymphomas and a pre-B cell leukemia.

tissues [268]. All other round cell tumors of childhood, i.e. Ewing's sarcoma, neuroblastoma, rhabdomyosarcoma [212, 269], and PNET [personal experience], including some lymphomas [270], are negative for LCA.

Other antibodies raised against surface determinants of hematopoietic cells, such as BA-1, BA-2, and FMG25, cross-react with tissues derived from neural crest, including neuroblastoma, and PNET [271, 272]. A weak reaction with Ewing's sarcoma is observed with BA-1 and BA-2, but not with FMG25. BA-1 and FMG25 do not react, but BA-2 reacts with rhabdomyosarcomas [273].

MB2, an antigen of B-cell lymphoma, does not stain neuroblastoma and rhabdomyosarcoma, except for differentiating ganglion cells and large rhabdomyoblasts [274]. Most Ewing's sarcomas and PNET are positive, suggesting that MB2 may distinguish Ewing's sarcoma/PNET from neuroblastoma and rhabdomyosarcoma [272].

Of other evaluated markers, HLA-DR was both positive [169] and negative in Ewing's sarcoma, but negative in neuroblastoma and rhabdomyosarcoma and positive in lymphoma [273]. Leu-M2 was positive in most Ewing's sarcomas and negative in rhabdomyosarcoma [169].

Another antibody originally raised against surface antigenic determinants of natural killer cells and subsequently found to be reactive in central and peripheral neuroectodermal and neuroendocrine tumors is HNK-1 (commercially available as Leu-7) [275]. HNK-1 recognizes epitopes on the myelin-associated glycoprotein (MAG) of central and peripheral nervous system [276], as well as glycoproteins that belong to the family of neural cell adhesion molecules (N-CAM) [277]. Positive staining with HNK-1 has been detected in neuroectodermal tissues and tumors, such as neuroblastoma, PNET, brain tumors, and carcinoids [275]. Ewing's sarcoma cell lines [168, 275] and tumors [184] and most mesenchymal chondrosarcomas [185] have also been reported positive. The surface antigen detected by the HNK-1 antibody in Ewing's sarcoma cell lines was found by thin layer chromatography to be an acidic glycolipid [168] with migration properties identical to those of the glycolipid detected by HNK-1 in the peripheral nervous system [278]. Although originally thought to stain only neuroblastoma, PNET and Ewing's sarcoma, we and others have found intense staining of rhabdomyosarcoma [279].

Neuroectodermal Antigens

HSAN 1.2 [280] and KP-NAC8 [281] have been raised against human neuroblastoma cell lines. Mab 5A7, an antibody raised against a cytoplasmic neuroblastoma protein, is the only antibody in the series not directed against surface antigenic determinants [282]. HSAN 1.2 and Mab 5A7 appear to be highly specific for neuroblastoma and PNET [14, 282, 283]. HSAN 1.2 shows strong immunoreactivity in neuroblastoma, weak immunoreactivity in PNET, and no reactivity in any other tumor, including Ewing's sarcoma, lymphoma, and rhabdomyosarcoma [14]. Two cell lines, originally considered as rhabdomyosarcoma, were subsequently characterized as PNET when stained with HSAN 1.2 [272], verifying its value in identifying neural tissues. Ewing's sarcoma cell lines reported to be positive with HSAN 1.2 [272] may reflect a neuronal stage of in vitro differentiation. Mab 5A7 was positive in neuroblastoma and PNET, and negative in Ewing's sarcoma and in cell lines from rhabdomyosarcoma and other tumors, including lymphoma [282]. Both HSAN 1.2 and Mab 5A7 work well on frozen, but not on paraffin-embedded tissues and are not currently commercially available. The KP-NAC8 antibodies cross-react with leukemia [283] and rhabdomyosarcoma [280].

UJ13A, UJ127.11 and UJ167.11 antibodies have been raised against human fetal brain cells. The UJ127.11 works in paraffin sections and stains neuroblastoma, PNET and astrocytoma but not rhabdomyosarcoma, Ewing's sarcoma and hematopoietic malignancies, except for leukemia [272, 283]. UJ13A and UJ167.11 stain neuroblastoma and PNET, do not cross-react with lymphoma, but stain rhabdomyosarcoma [280]. The antigen recognized by the UJ13A antibody is the neural cell adhesion molecule (N-CAM) [284], which is widely expressed in neuroendocrine tissues and neoplasms [285] and in myogenous cells [279]. Because of the localization of the N-CAM gene to chromosome 11 close to the breakpoint of the specific t(11:22) translocation [286], and the differences in sialylation and expression of the N-CAM molecule between neuroblastoma and Ewing's sarcoma, it was thought that N-CAM gene may be altered in Ewing's sarcoma/PNET, but preliminary DNA analysis has not revealed any cross rearrangement [287].

Antigens Distinguishing between Ewing's Sarcoma/PNET and Neuroblastoma

W6/32, a monoclonal antibody used to distinguish neuroblastoma from PNET [14], was raised against native HLA-A,B,C (class I) molecules [288]. More extensive studies, however, have been undertaken with antibodies against β_2-microglobulin, the invariant chain of the wider family of class I histocompatibility molecules, which is associated with HLA-A,B,C molecules [289]. Major histocompatibility complex class I antigens are expressed in most normal tissues and tumors [290], including PNET and olfactory neuroblastoma [38, 198], but not conventional neuroblastoma in vivo [291] or in vitro [292]. In spite of its absence from neuroblastoma by immunocytochemical methods, a β_2-microglobulin-like chain was found to be synthesized by neuroblastoma cells in vitro [292] and could be induced in vitro [293] and in vivo [294] with γ-interferon. High levels of β_2-microglobulin are expressed in stage IV-S neuroblastoma and in neuroblastoma composed predominantly of ganglion cells [295], suggesting a biologic role for β_2-microglobulin in neuroblastoma. Since HLA class I surface antigens are responsible for recognition of self from nonself and mediate the lytic function of cytotoxic T lymphocytes [296], β_2-microglobulin, in association with some other tumor-specific antigens, may be responsible for tumor cell lysis by cytotoxic T cells, whereas low expression may facilitate tumor cell resistance to lysis [295]. β_2-Microglobulin concentration is reduced in a few aggressive neoplasms [297, 298], and in rat neuroblastoma after N-myc amplification, which also increases neoplastic aggressiveness [299]. Conversely, high levels of β_2-microglobulin in stage IV-S neuroblastoma encourage tumor cell lysis by cytotoxic lymphocytes, allowing for complete regression later in infancy on full development of the immune system [295]. Localization of β_2-microglobulin in ganglion cells within otherwise negative neuroblastomas [198, 295] may be explained if β_2-microglobulin is a differentiation marker in neuroblastoma, or if ganglion cells, as terminally differentiated cells, lack tumor-specific antigens and are not susceptible to cytotoxic T-cell lysis.

HBA.71, raised against a PNET cell line and recognizing a cell surface antigen of Ewing's sarcoma/PNET-derived cells [167], can be used to distinguish neuroblastoma from PNET/Ewing's sarcoma and probably rhabdomyosarcoma. The HBA.71 antigen, although originally thought to be related to the Thy-1 antigen [300], has been found identical or closely related to the product of the pseudoautosomal gene MIC2, which had been previously identified by the monoclonal antibody 12E7 in hematopoietic cells [301–303]. This antigen has so far been detected in Ewing's sarcoma and PNET cell lines and tumors, in astrocytomas, some neuroendocrine tumors, lymphomas and sarcomas, but not in neuroblastoma. Only a few rhabdomyosarcomas have shown a small subset of positive tumor cells [302, 303]. Using the 12E7 antibody, we found intense positivity of Ewing's sarcoma/PNET and negative staining of most rhabdomyosarcomas, except for larger myoblastic cells [unpubl. data]. This antibody, which works well in formalin-fixed, paraffin-embedded tissues, may be a very useful marker in the distinction of PNET/Ewing's sarcoma from neuroblastoma and rhabdomyosarcoma.

Endothelial Cell Markers

The endothelial cell markers factor VIII-related antigen, and *Ulex europaeus* I lectin (UEA I) were negative in Ewing's sarcoma of bone and soft tissue [170], arguing against an endothelial cell origin of the tumor. However, these findings are not absolutely conclusive, since factor VIII-related antigen is usually absent from malignant endothelial cells, and UEA I may or may not be present in benign and malignant endothelial cell tumors.

Cytogenetics and Molecular Genetics

Neuroblastoma

Tumor suppressor genes had been hypothesized and recently verified in various childhood cancers. Although a tumor suppressor gene has been hypothesized for neuroblastoma, it has not yet been found [304]. However, tumor suppressor genes have been associated with chromosomal deletions in childhood tumors [305, 306], and 70–80% of all neuroblastoma tumors and cell lines exhibit a consistent deletion of the distal portion of short arm of chromosome 1, i.e., a partial 1p monosomy [307, 308], thought to represent loss of a putative neuroblastoma (Nb) gene. Restriction fragment length polymorphism (RFLP) analysis revealed loss of heterozygosity in 28% of neuroblastomas at the distal end of the short arm of chromosome 1, in the area between 1p36.1 and 1p36.3, which could be the neuroblastoma locus [309]. A higher frequency of chromosome 1 deletion (70%) over loss of heterozygosity (28%) may be explained in several ways, but the association of lost heterozygosity with high stages of disease [309] suggests that it plays an important role in tumor progression as a secondary event in the course of clonal evolution [304]. Since chromosome 1p deletion syndromes have not been reported, deletions sufficiently large as to be visible cytogenetically may be incompatible with life [304]. Although abnormalities of chromosome 1 have been reported in various other tumors, they usually consist of trisomy for the long arm and not monosomy for the short arm, which is largely restricted to neuroblastoma and some neuroectodermal tumors [310].

Other cytogenetic abnormalities in neuroblastoma consist of extrachromosomal double minutes in about a third of primary tumors and half of neuroblastoma cell lines, and chromosomally integrated homogeneously staining regions (HSRs) [307, 308]. A switch from double minutes to HSRs was noted in one neuroblastoma cell line established from recurrent tumor therapy [311]. Other abnormalities reported with significant frequency in approximately 20% of neuroblastomas involved trisomies for the long arms of chromosomes 1 and 17 [308].

N-myc Oncogene

Double minutes and HSRs have been attributed to cytogenetic manifestations of gene amplification [307, 312], but the nature of the amplified gene was unknown until the discovery of the amplified N-myc oncogene in neuroblastoma cell lines [313–315]. The normal single copy locus of this oncogene was mapped to the distal short arm of chromosome 2, and the amplified N-myc gene to the HSRs of various chromosomes [313, 316]. Amplification of a large region of chromosome 2 occurs initially as extrachromosomal double minutes [317]. Double minutes and HSRs contain multiple copies of N-myc [318].

N-myc amplification occurred in approximately 38% of neuroblastomas [319] and was exclusively associated with high stage of disease. N-myc amplification also associates with rapid tumor progression and fatal outcome regardless of the stage of the tumor [320, 321]. Five to 10% of low stage (I, II and IV-S) neuroblastic tumors have N-myc amplification [322–325]. In some studies [324, 325], N-myc amplification was associated only with the adrenal neuroblastoma, suggesting a different mechanism of tumor progression in adrenal and extraadrenal neuroblastoma. N-myc amplification was detected successfully by dot-blot hybridization of DNA extracted from formalin-fixed paraffin-embedded archival tissues [326] and by in situ hybridization with a single-step silver enhancement technique, which allowed correlation between N-myc copy numbers and histopathology [327].

There is a statistically significant correlation between N-myc amplification and loss of heterozygosity in chromosome 1p [328]. Both N-myc amplification [319–323, 328] and chromosome 1p deletion [329, 330] predict a poor clinical outcome. Although it is not clear if N-myc amplification and chromosome 1 deletions are independent prognostic parameters, they appear to define two genetically distinct groups of neuroblastoma when combined with flow cytometric data [331]. The first subcategory of neuroblastoma has hyperdiploid or triploid modal karyotypes, which exhibit a few, if any, structural chromosomal abnormalities, lack N-myc amplification, and occur in patients younger than 1 year with low-stage disease. The second subcategory encompassed neuroblastomas with a near-diploid or tetraploid mode, exhibiting numerous structural chromosomal abnormalities and N-myc amplification and occurring in patients older than 1 year and presenting with stage III or IV disease.

N-myc amplification also correlates with significantly decreased amounts of vanillylmandelic acid and homovanillic acid in the tumor, but not with their ratio in the urine [323].

In general, N-myc copy numbers correlate with levels of expression [311, 332–336], although high levels of N-myc expression may be associated with single N-myc copy numbers and low levels in the absence of N-myc amplification have been observed [324, 334, 336]. Southern blot analysis of frozen tumor tissues has shown stable N-myc copy numbers in simultaneous or consecutive tumor samples obtained at different times during treatment, suggesting that N-myc amplification is an intrinsic biological property in neuroblastoma [326, 328].

Although the relationship between N-myc expression and disease progression is still controversial [314, 324, 332–336], employment of in situ hybridization techniques in several studies has shown that high N-myc RNA levels correspond to amplified N-myc sequences and correlate inversely with histopathologic differentiation and prognosis [332, 334, 337]. Tumors with terminal or partial differentiation to ganglioneuroma are negative for N-myc expression [324, 336]. Characterization of the N-myc gene product, a pair of 65- and 67-kd DNA-binding phosphoproteins, and development of polyclonal and monoclonal antibodies against this product have offered a new way of studying the expression of the N-myc gene in tissue sections employing standard immunohistochemical methods, although cross-reactivity of the antibody with other tissues or lack of expression in spite of gene amplification have been reported [338–341].

Although amplification and overexpression of the N-myc gene were originally thought to be properties of neuroectodermal cells, predominantly of neuroblastoma, small cell carcinoma of the lung and retinoblastoma [313–315, 318–325, 342, 343], they have been additionally reported in Wilms' tumor [344, 345], teratocarcinoma stem cells [346], astrocytoma [347], C-cell neoplasia of the thyroid [348], and sporadic rhabdomyosarcomas [103, 349–352]. N-myc amplification or overexpression was absent from other rhabdomyosarcoma tissues and cell lines [326, 353]. N-myc amplification was detected in two central (cerebellar) PNET with a differentiated (neuronal) phenotype [354], in contrast to its association with a poorly differentiated phenotype in neuroblastoma [327].

The selective pressures that lead to high levels of N-myc amplification in neuroblastoma are unknown, as is the precise function of the N-myc oncogene product. Although c-myc regulates cell growth [355], N-myc and c-myc are evidently regulated separately [356]. Therefore, growth arrest by depletion of the intracellular polyamine reservoir or by serum deprivation does not impede expression of N-myc in neuroblastoma cells [357]. Using somatic cell hybrid studies, hybrids derived from neuroblastoma or retinoblastoma cells with amplified N-myc and from HeLa cells devoid of N-myc amplification are nontumorigenic and lack N-myc expression, implying a strong selective pressure against N-myc amplification. When tumorigenicity is reexpressed in the hybrid cells, N-myc mRNA levels are not increased, suggesting that N-myc overexpression is not necessary for reexpression of the tumorigenic potential [358].

N-myc, in contrast to c-myc, is expressed in high levels in the embryo, and is restricted to certain tissues and stages of development in the mouse embryo [346] and fetal human brain [359]. Specifically, appreciable N-myc is expressed in immature neural cells, disappears with differentiation, and may be unrelated to cell proliferation, because high levels are seen in the primordial cortex which grows by accretion and not by cell division. N-myc may, therefore, relate to cell differentiation rather than proliferation. The level of N-myc expression in neuroblastoma cells decreases dramatically following differentiation in vitro, independently of cell growth [357, 360, 361]. Although rapid progression of neuroblastomas containing amplified N-myc gene suggests selective growth advantage of these tumor cells, transfection experiments with N-myc have shown no differences in the growth rates of N-myc-carrying neuroblastoma cells versus control transfectants [195]. Instead, high levels of N-myc gene in transfected neuroblastoma cells prevent enhancement of neurofilament gene expression after treatment with retinoic acid, suggesting that N-myc may inhibit differentiation by affecting the expression of maturation-related genes [195]. The decreased N-myc transcription by retinoic acid has led to the postulation that migrating neuroblasts, upon arrival in the adrenal cortex, receive signals from naturally occurring morphogens to proceed along a normal, neuronal phenotype, and that disturbance of this developmental process may lead to carcinogenesis [362]. After transfection of rat neuroblastoma cells with an N-myc expression vector, increased metastatic potential and decreased neural cell adhesion molecule (N-CAM) expression appeared, suggesting an indirect role of N-myc in the metastatic potential of neuroblastoma via pathways controlling cell adhesion molecules [363].

C-myc

C-myc, another oncogene of the myc family, is never amplified in neuroblastoma [324, 326], unlike PNET and rhabdomyosarcoma [326, 353]. Most neuroblastomas exhibit transcription of N-myc and c-myc at the same time, except for those with amplified or highly expressed N-myc which lack c-myc expression, and those with complete or partial differentiation to ganglioneuroma, which lack N-myc expression [324]. Experimentally, endogenous c-myc expression can be suppressed by high levels of N-myc expression [364]. Phorbol ester-induced maturation of a human neuroblastoma cell line was accompanied by down-regulation of c-myc and N-myc expression, suggesting that a transient c-myc decline may be important in the maturation process of neuroblastoma, like the decline of N-myc [357].

Other Oncogenes

A few other oncogenes are activated in neuroblastoma. The N-ras oncogene is the transforming gene of the extraadrenal human neuroblastoma cell line SK-N-SH [365, 366]. Although ras gene mutations had not been identified in neuroblastoma [367], using the polymerase chain reaction, activated N-ras genes were shown in 2 of 19 neuroblastomas [368], both of which were of low stage (I and II respectively), suggesting a possible relationship between N-ras gene mutation and early stages of neuroblastoma. Similarly, the Ha-ras gene product correlated with a favorable prognosis and early stages of disease at diagnosis [369].

A relatively high expression of c-ets oncogene, plus high levels of tyrosine kinase activity (pp60^{C-src}) in neuroblastoma suggest posttranslationally activated c-src gene [370]. Enhancement of c-src expression during neuronal differentiation and suppression during Schwannian differentiation of neuroblastoma cells in vitro have suggested a relationship of c-src expression to neuronal differentiation in vivo [371].

Multidrug Resistance Gene

The double minutes and HSRs in neuroblastoma led to a search for amplified sequences, resulting in the identification of the amplified N-myc oncogene. One of the initially sought, but not yet successfully achieved correlations, involved the multidrug resistance (mdr) gene [304]. One hypothetical mechanism of tumor progression and drug resistance is the selection of tumor cells with amplified mdr genes occurring in tumors with double minute chromosomes [372].

Studies concerning expression of mdr-gene and its product P-glycoprotein in neuroblastoma have shown that mdr-1/P-glycoprotein expression: (1) increases after retinoic acid-induced differentiation of neuroblastoma cell lines, although it is not necessarily associated with increased cytotoxic drug accumulation [373]; (2) is higher in more differentiated tumors as determined by in situ hybridization and immunoperoxidase techniques [374]; (3) correlates inversely with N-myc expression [375] suggesting an indirect relationship with differentiation, and (4) occurs at highest levels in neuroblastoma exposed to chemotherapy [376] suggesting that either chemotherapy induces differentiation, and thus more differentiated cells express higher levels of mdr-1, or that it results in selection of mdr-1/P-glycoprotein-expressing and hence more resistant cell types. However, the functional role of mdr-gene and P-glycoprotein in neuroblastoma is still unknown, and more than one event may be required to establish a functional P-glycoprotein phenotype.

Peripheral PNET and Ewing's Sarcoma

A specific chromosomal translocation, t(11;22)(q24;q11), originally described in Ewing's sarcoma [29], has been a consistent finding in a large proportion of cell lines and primary tumors of the Ewing's sarcoma-PNET family, including the extraskeletal Ewing's sarcoma and the small round cell tumor of the thoracopulmonary region (Askin tumor) [16, 29, 30, 377, 378]. Chromosome markers carrying breakpoints at 22q12 or 11q24 occurred in most cases of Ewing's sarcoma usually as *standard*: the simple reciprocal t(11;22)(q24;q12) translocation, less often as *complex*: translocations involving chromosomes other than 22 and 11, or *variant*: translocations involving chromosome 22 and chromosome(s) other than 11 [377]. A small proportion of cases had no rearrangement of chromosome 11q24 or 22q12, although one case had an isochromosome 11, i(11q) [16]. The occurrence of alterations in chromosome 22 in 92% of cases of Ewing's sarcoma has led to the suggestion that chromosome 22 may be more important than chromosome 11, and that the microscopically detected 11;22 translocation may not be central in the pathogenesis of Ewing's sarcoma [379]. Both chromosomes are also involved in a constitutional t(11;22) translocation, but chromosomal in situ hybridization data suggest that the breakpoints in the Ewing's sarcoma-associated translocation are different from those in the constitutional translocation [380].

This highly consistent reciprocal translocation has been considered a reliable marker for Ewing's sarcoma, peripheral PNET and Askin tumor, suggesting a common histogenesis for these tumors and a potential diagnostic marker to distinguish them from other small round cell tumors in children, especially neuroblastoma. Molecular genetics is only rarely needed for distinguishing neuroblastoma from peripheral PNET, since neuroblastoma usually shows sufficient morphologic differentiation by light and electron microscopy, lacks expression of β_2-microglobulin and HBA-71 antigen, and exhibits a characteristic clinical picture. On the other hand, distinction of peripheral PNET from neuroblastoma is not trivial, since these tumors, in spite of a shared morphology, have great differences in their biology and clinical behaviors.

Other tumors of neuroectodermal origin, such as esthesioneuroblastoma [38, 381] and neuroendocrine tumor of the small intestine [382], have had 11;22 translocations. We have also observed a t(11;22)(q24;q12) translocation in an ectomesenchymoma cell line and in a cell clone of a conventional alveolar rhabdomyosarcoma that subsequently expressed a focal PNET phenotype [unpubl. data]. These findings suggest that the cytogenetic abnormalities in a given tumor represent the existing cell clones and may occasionally help identify cell clones or pathways of differentiation that are morphologically obscure.

The role of t(11;22) is unclear. Neither the c-sis oncogene on chromosome 22q [383], nor the c-ets-1 oncogene on chromosome 11q [380] appears to be implicated. The c-sis translocates from chromosome 22 to 11, but is not activated or rearranged in Ewing's sarcoma [384, 385]. PNET and neuroblastoma cell lines do not express detectable c-sis transcripts [386]. Similarly, c-ets shows no rearrangement in Ewing's sarcoma and shows variable but minimal patterns of expression in PNET and Ewing's sarcoma cell lines compared to those of neuroblastoma cell lines [31, 34].

Molecular genetic studies involving amplification and expression of several other protooncogenes have shown that N-myc is not amplified in any of the peripheral PNET or Ewing's sarcoma cell lines, in contrast to classic neuroblastoma, although low levels are expressed [31, 34]. All studied peripheral PNET and Ewing's sarcoma cell lines exhibit high levels of c-myc RNA, which are not necessarily associated with c-myc amplification [332]. Indistinguishable patterns of protooncogenes are expressed in peripheral PNET and Ewing's sarcoma [34]. Tyrosine kinase activity of c-src oncogene is high in PNET and Ewing's sarcoma as in neuroblastoma [387]. Another oncogene, dbl, was consistently expressed and initially considered to be a marker for Ewing's sarcoma [388], but subsequent studies have shown dbl expression in other tumors as well [Navarro et al., unpubl. data, Vechio and Eva, pers. commun.]. Other oncogenes expressed in relatively high levels in PNET/Ewing's sarcoma, as in neuroblastoma cell lines, are the c-myb and c-raf oncogenes [31, 32, 386, 389].

Flow Cytometry

Although flow cytometric studies of peripheral PNET and Ewing's sarcoma are limited [390], DNA ploidy and proliferative activity in childhood neuroblastoma has been correlated with clinical stage, histologic subtype, and the level of N-myc amplification [391–394].

Cytometric studies of numerous neoplasms have associated abnormal DNA content with poor prognosis, with the notable exceptions of neuroblastoma [391, 392, 395] and acute lymphoblastic leukemia [396]. Unresectable hyperdiploid neuroblastoma responded better to chemotherapy than did DNA-diploid neuroblastoma [392]. A favorable outcome, slow growth rate, low clinical stage, and a favorable histopathologic pattern was associated with aneuploid neuroblastoma [391, 394, 395]. In contrast, DNA diploidy correlated with unfavorable histology, and high proliferative activity [395]. Most patients with aneuploid neuroblastoma and good clinical outcome were usually younger than 18 months [391, 393, 395]. Clonal hyperdiploid abnormalities were also found in stage IV-S neuroblastoma and supported malignant transformation over benign hyperplasia [392]. Retrospective flow cytometric evaluation of neuroblastoma and ganglioneuroma tumor samples has shown that nuclear DNA content is a stable tumor marker, remaining the same at different times during the disease [394].

In relation to catecholamine secretion profiles, diploid neuroblastomas, in contrast to aneuploid ones, secrete higher levels of early pathway metabolites, such as 3,4-dihydroxyphenylalanine (DOPA), dopamine, and homovanillic acid and are more likely to be in stages II and IVs [397]. The flow cytometric data in neuroblastoma agree with the cytogenetic data, both showing hyperdiploid karyotypes in the triploid range associated with a favorable outcome in children less than 1 year of age and favorable stage of disease [329–331], and diploid or pseudodiploid karyotypes associated with advanced stage in patients older than 1 year, progressive disease, deletions of chromosome 1, double minutes, and N-myc amplification [329–331, 395, 396]. DNA ploidy and proliferative activity evaluated by flow cytometry, may therefore provide additional prognostic information in neuroblastoma. The urinary catechol-amine screening program that has been underway in Japan for over a decade identified aneuploid subgroups of neuroblastoma at low stages of disease [330], suggesting either evolution of neuroblastoma from a more favorable to a more unfavorable genotype and phenotype later in life, or two distinct subsets of neuroblastomas, with the more favorable presenting earlier in life and detected by the catecholamine screening methods [329–331].

Summary and Conclusions

In conclusion, the group of peripheral primitive neuroectodermal tu-mors has been redefined in recent years on the basis of cytogenetic, molecular genetic and more precisely defined histopathologic characteristics. Although in the past, many tumors had been called Ewing's sarcoma, currently this diagnosis is limited to tumors which cannot be more specifically classified on the basis of their ultrastructural and immunophenotypic characteristics. Most small round cell tumors previously classified as Ewing's sarcoma are now classified as peripheral PNET. The consistent cytogenetic abnormality in Ewing's sarcoma and peripheral PNET and patterns of neurotransmitter enzymes have supported a common neuroectodermal origin.

The precise characterization of soft tissue Ewing's sarcoma is further complicated by the several primitive rhabdomyosarcomas that may exhibit a similar light microscopic appearance. The importance of histopathologic distinction among these various round cell tumors of childhood is well recognized. Furthermore, primitive tumors with overlapping neural and mesenchymal features, known as malignant ectomesenchymoma, are now identified more often than previously.

Finally, molecular biologic and cytogenetic differences between periph-eral PNET and neuroblastoma have confirmed their clinical and biologic differences, in spite of their morphologic similarities. Molecular genetic and flow cytometric evaluation have contributed to the distinction of groups with prognostic significance and offer possibilities for new clinical trials.

References

1 Dehner LP: Peripheral and central primitive neuroectodermal tumors: A nosologic concept seeking a consensus (editorial). Arch Pathol Lab Med 1986;110:997–1004.

2   Triche TJ: Neuroblastoma – biology confronts nosology (editorial). Arch Pathol Lab
    Med 1986;110:994–996.
3   Triche TJ: Neuroblastoma and other childhood neural tumors. A review. Pediatr
    Pathol 1990;10:175–193.
4   Dehner LP: Will the 'real' neuroblastoma please stand up? (letter) Arch Pathol Lab
    Med 1985;109:794–795.
5   Stout AP: A tumor of the ulnar nerve. Proc NY Pathol Soc 1918;18:2–12.
6   Nesbitt KA, Vidone RA: Primitive neuroectodermal tumor (neuroblastoma) arising
    in sciatic nerve of a child. Cancer 1976;37:1562–1570.
7   Aleshire SL, Glick AD, Cruz V, et al: Neuroblastoma in adults. Pathologic findings
    and clinical outcome. Arch Pathol Lab Med 1985;109:352–356.
8   Hashimoto H, Enjoji M, Nakajima T, et al: Malignant neuroepithelioma (peripheral
    neuroblastoma). Am J Surg Pathol 1983;7:309–318.
9   Llombart-Bosch A, Lacombe MJ, Peydro-Olaya A, et al: Malignant peripheral
    neuroectodermal tumours of bone other than Askin's neoplasm: characterization of
    14 new cases with immunohistochemistry and electron microscopy. Virchows Arch
    [A] 1988;412:421–430.
10 Llombart-Bosch A, Terrier-Lacombe MJ, Peydro-Olaya A, et al: Peripheral neuroec-
    todermal sarcoma of soft tissue (peripheral neuroepithelioma): A pathologic study of
    ten cases with differential diagnosis regarding other small, round-cell sarcomas. Hum
    Pathol 1989;20:273–289.
11 Marina NM, Etcubanas E, Parham DM, et al: Peripheral primitive neuroectoder-
    mal tumor (peripheral neuroepithelioma) in children. Cancer 1989;64:1952–
    1960.
12 Schmidt D, Harms D, Burdach S: Malignant peripheral neuroectodermal tumours of
    childhood and adolescence. Virchows Arch [A] 1985;406:351–365.
13 Schmidt D, Harms D, Jurgens H: Malignant peripheral neuroectodermal tumors.
    Histological and immunohistological conditions in 41 cases. Zentralbl Allg Pathol
    1989;135:257–268.
14 Tsokos M, Donner L, Reynolds CP, et al: Peripheral neuroepithelioma: An entity
    distinct from classic neuroblastoma (abstract). Lab Invest 1985;52:70A.
15 Bolen JW, Thorning D: Peripheral neuroepithelioma: A light and electron micro-
    scopic study. Cancer 1980;46:2456–2462.
16 Haas OA, Chott A, Ladenstein R, et al: Poorly differentiated, neuron-specific
    enolase-positive round cell tumor with two translocations t(11;22) and t(21;22).
    Cancer 1987;60:2219–2223.
17 Harper PG, Pringle J, Souhami RL: Neuroepithelioma: A rare malignant peripheral
    nerve tumor of primitive origin. Cancer 1981;48:2282–2287.
18 Mackay B, Luna MA, Butler JJ: Adult neuroblastoma. Electron microscopic obser-
    vations in nine cases. Cancer 1976;37:1334–1351.
19 Mackay B, Ordonez NG: Adult neuroblastoma of bone: a case report. Ultrastruct
    Pathol 1987;114:455–464.
20 Pysher TJ, Boyer RS, Walker ML: Case 3. Primitive neuroectodermal tumor –
    Peripheral neuroepithelioma. Pediatr Pathol 1989;9:185–191.
21 Samuel AW: Primitive neuroectodermal tumor arising in the ulnar nerve. A case
    report. Clin Orthop 1982;167:236–328.
22 Voss BL, Pysher TJ, Humphrey GB: Peripheral neuroepithelioma in childhood.
    Cancer 1984;54:3059–3064.

23 Jaffe R, Santamaria M, Yunis EJ, et al: The neuroectodermal tumor of bone. Am J Surg Pathol 1984;8:885–898.

24 Askin FB, Rosai J, Sibley PK, et al: Malignant small cell tumor of the thoracopulmonary region in childhood. Cancer 1979;43:2438–2451.

25 Gonzalez-Crussi F, Wolfson SL, Misugi K, et al: Peripheral neuroectodermal tumors of the chest wall in childhood. Hum Pathol 1984;54:2519–2527.

26 Linnoila RI, Tsokos M, Triche TJ, et al: Evidence for neural origin and PAS-positive variants of the malignant small cell tumor of thoracopulmonary region ('Askin tumor'). Am J Surg Pathol 1986;10:124–133.

27 Llombart-Bosch A, Lacombe MJ, Contesso G, et al: Small round blue cell sarcoma of bone mimicking atypical Ewing's sarcoma with neuroectodermal features. Cancer 1987;60:1570–1582.

28 Schmidt D, Mackay B, Ayala AG: Ewing's sarcoma with neuroblastoma-like features. Ultrastruct Pathol 1982;3:143–151.

29 Turc-Carel C, Aurias A, Mugneret F, et al: Chromosome study of Ewing's sarcoma (ES) cell lines. Consistency of a reciprocal translocation t(11;22)(q24;q12). Cancer Genet Cytogenet 1984;12:1–19.

30 Whang-Peng J, Triche TJ, Knutsen T, et al: Chromosome translocation in peripheral neuroepithelioma. N Engl J Med 1984;311:584–585.

31 Thiele CJ, McKeon C, Triche TJ, et al: Differential protooncogene expression characterizes histopathologically indistinguishable tumors of the peripheral nervous system. J Clin Invest 1987;80:804–811.

32 Tsokos M, Pfeifer A, Ross RA, et al: Peripheral neuroepithelioma versus neuroblastoma in vitro: Morphology, enzyme analysis, and oncogene expression (abstract). Lab Invest 1987;56:80A.

33 Cavazzana AO, Miser JS, Jefferson J: Experimental evidence for a neural origin of Ewing's sarcoma of bone. Am J Pathol 1987;127:507–518.

34 McKeon C, Thiele CJ, Ross RA, et al: Indistinguishable patterns of protooncogene expression in two distinct but closely related tumors: Ewing's sarcoma and neuroepithelioma. Cancer Res 1988;48:4307–4311.

35 Jacobson SA: Polyhistioma: A malignant tumor of bone and extraskeletal tissues. Cancer 1977;40:2116–2130.

36 Hutter RVP, Foote FW Jr, Frances KC, et al: Primitive multipotential primary sarcoma of bone. Cancer 1965;19:1–25.

37 Tsokos M, Miser J, Horowitz M, et al: Primitive round cell sarcoma of bone: a group of tumors resembling Ewing's sarcoma by light microscopy, but with distinctive ultrastructural appearance (abstract). Lab Invest 1988;58:96A.

38 Cavazzana AO, Navarro S, Noguera R, et al: Olfactory neuroblastoma is not a neuroblastoma but is related to primitive neuroectodermal tumor (PNET). Prog Clin Biol Res 1988;271:463–474.

39 Choi HS, Anderson PJ: Olfactory neuroblastoma: an immunoelectron microscopic study of S-100 protein-positive cells. J Neuropathol Exp Neurol 1986;4:576–587.

40 Cullen MJ, Cusak DA, O'Brian DS, et al: Neurosecretion of arginine vasopressin by an olfactory neuroblastoma causing reversible syndrome of antidiuresis. Am J Med 1986;81:911–916.

41 Llombart-Bosch A, Carda C, Peydro-Olaya A, et al: Pigmented esthesioneuroblastoma showing dual differentiation following transplantation in nude mice. An

immunohistochemical, electron microscopical, and cytogenetic analysis. Virchows Arch [A] 1989;414:199–208.

42 Micheau C: A new histochemical and biochemical approach to olfactory esthesioneuroblastoma; a nasal tumor of neural crest origin. Cancer 1977;40:314–318.

43 Micheau C, Guerinot F, Bohuon C, et al: Dopamine-β-hydroxylase and catecholamines in an olfactory esthesioneuroma. Cancer 1975;35:1309–1312.

44 Burget EO, Nesbit ME, Garnsey LA, et al: Multimodal therapy for the management of nonpelvic, localized Ewing's sarcoma of bone: Intergroup study IESS-II. J Clin Oncol 1990;8:1514–1524.

45 Jürgens H, Exner U, Gadner H, et al: Multidisciplinary treatment of primary Ewing's sarcoma of bone. A 6-year experience of a European cooperative trial. Cancer 1988;61:23–32.

46 Miser JS, Kinsella TJ, Triche TJ, et al: Treatment of Ewing's sarcoma of bone in children and young adults: six months of intensive combined modality therapy without maintenance. J Clin Oncol 1988;6:484–490.

47 Bader JL, Horowitz ME, Dewan R, et al: Intensive combined modality therapy of small round cell and undifferentiated sarcomas in children and young adults: local control and patterns of failure. Radiother Oncol 1989;16:189–201.

48 Kajanti M: Neuroblastoma in 88 children. Clinical features, prognostic factors, results, and late effects of therapy. Ann Clin Res 1983;15(suppl 39):1–68.

49 Seeger RC, Siegel SE, Sidell N: Neuroblastoma: Clinical perspectives, monoclonal antibodies, and retinoic acid. Ann Intern Med 1982;97:873–884.

50 Triche TJ, Askin FB, Kissane JM: Neuroblastoma, Ewing's sarcoma and the differential diagnosis of small-, round-, blue-cell tumors; in Finegold M (ed): Pathology of Neoplasia in Children and Adolescents. Philadelphia, Saunders, 1986, pp 145–195.

51 Bolande RP: Spontaneous regression and cytodifferentiation of cancer in early life: the oncogenic grace period. Surv Synth Pathol Res 1985;44:296–311.

52 Garvin JH, Lack EE, Berenberg W, et al: Ganglioneuroma presenting with differentiated skeletal metastases. Report of a case. Cancer 1984;54:357–360.

53 Koppersmith DL, Powers JM, Hennigar GR: Angiomatoid neuroblastoma with cytoplasmic glycogen. A case report and histogenetic considerations. Cancer 1980;45:553–560.

54 Cozzutto C, Carbone A: Pleomorphic (anaplastic) neuroblastoma. Arch Pathol Lab Med 1988;112:621–625.

55 Gonzalez-Crussi F, Hsueh W: Bilateral adrenal ganglioneuroblastoma with neuromelanin. Clinical and pathologic observations. Cancer 1988;61:1159–1166.

56 Mullins JD: A pigmented differentiating neuroblastoma: A light and ultrastructural study. Cancer 1980;46:522–528.

57 Triche TJ, Ross WE: Glycogen-containing neuroblastoma with clinical and histopathologic features of Ewing's sarcoma. Cancer 1978;41:425–432.

58 Yunis EJ, Agostini RM, Walpusk JA, et al: Glycogen in neuroblastomas. A light- and electron-microscopic study of 40 cases. Am J Surg Pathol 1979;3:313–323.

59 Mierau GW, Berry PJ, Orsini EN: Small round cell neoplasms: Can electron microscopy and immunohistochemical studies accurately classify them? Ultrastruct Pathol 1985;9:99–111.

60 Romansky SG, Crocker DW, Shaw KNF: Ultrastructural studies on neuroblastoma: Evaluation of cytodifferentiation and correlation of morphology and biochemical survival data. Cancer 1978;42:2392–2398.

61 Taxy JB: Electron microscopy in the diagnosis of neuroblastoma. Arch Pathol Lab Med 1980;104:355–360.

62 Luse SA: Synaptic structures occurring in a neuroblastoma. Arch Neurol 1964;11: 185–190.

63 Gonzalez-Angulo A, Reyes HA, Reyna AN: The ultrastructure of ganglioneuroblastoma: Observations on neoplastic ganglion cells. Neurology 1965;15:242–252.

64 Staley NA, Polesky HF, Bensch KG: Fine structural and biochemical studies on the malignant ganglioneuroma. J Neuropathol Exp Neurol 1967;26:634–653.

65 Jaffe N: Neuroblastoma: Review of the literature and an examination of factors contributing to its enigmatic character. Cancer Treat Rev 1976;3:61–82.

66 Beckwith JB, Martin RF: Observations on the histopathology of neuroblastomas. J Pediatr Surg 1968;3:106–110.

67 Sandstedt B, Jereb B, Eklund G: Prognostic factors in neuroblastomas. Acta Pathol Microbiol Immunol Scand [A] 1983;91:365–371.

68 Gitlow SE, Bertani Dziedzic L, Strauss L, et al: Biochemical and histologic determinants in the prognosis of neuroblastoma. Cancer 1973;4:898–905.

69 Makinen J: Microscopic patterns as a guide to prognosis of neuroblastoma in childhood. Cancer 1972;29:1637–1646.

70 Shimada H, Chatten J, Newton WA, et al: Histopathologic prognostic factors in neuroblastic tumors: definition of subtypes of ganglioneuroblastoma and an age-linked classification of neuroblastomas. J Natl Cancer Inst 1984;73:405–416.

71 Chatten J, Shimada H, Sather HN, et al: Prognostic value of histopathology in advanced neuroblastoma: A report from the Childrens' Cancer Study Group. Hum Pathol 1988;19:1187–1198.

72 Brook FB, Raafat F, Mann JR: Histologic and immunohistochemical investigation of neuroblastomas and correlation with prognosis. Hum Pathol 1988;19:879–888.

73 O'Neil JA, Littman P, Blitzer P, et al: The role of surgery in localized neuroblastoma. J Pediatr Surg 1985;20:708–712.

74 Evans AE, D'Angio G, Propert K, et al: Prognostic factors in neuroblastoma. Cancer 1987;59:1853–1859.

75 Joshi V, Altshuler G, Cantor A, et al: Prognostic significance of histopathologic features of neuroblastoma (NB): A grading system based on review of 211 cases from pediatric oncology group (POG) (abstract). Mod Pathol 1991;4:6P(32).

76 Ewing J: Diffuse endothelioma of bone. Proc NY Pathol Soc 1921;21:17–24.

77 Erladson RA: The ultrastructural distinction between rhabdomyosarcoma and the undifferentiated 'sarcomas'. Ultrastruct Pathol 1987;11:83–101.

78 Katenkamp D, Gudziol H, Kuttner K, et al: Olfactory neuroblastoma. Clinical course, light microscopic and ultrastructural findings in 3 cases. Zentralbl Allg Pathol 1986;132:57–70.

79 Azar HA, Jaffe ES, Berard CW, et al: Diffuse large cell lymphomas (reticulum cell sarcomas, histiocytic lymphomas): correlation of morphologic features with functional markers. Cancer 1980;46:1428–1441.

80 Kissane J, Askin F, Foulkes M, et al: Ewing's sarcoma of bone; clinicopathologic aspects of 303 cases from the Intergroup Ewing's sarcoma study. Hum Pathol 1983; 14:773–779.

81 Llombart-Bosch A, Contesso G, Henry-Amar M, et al: Histopathological predictive factors in Ewing's sarcoma of bone and its histopathological correlations: A retrospective study of 261 cases. Virchows Arch [A] 1986;409:627–640.

82 Mahoney JP, Alexander RW: Ewing's sarcoma. A light- and electron-microscopic study of 21 cases. Am J Surg Pathol 1978;2:283–298.

83 Navas-Palacios JJ, Aparicio-Duque R, Valdes MD: On the histogenesis of Ewing's sarcoma. An ultrastructural, immunohistochemical, and cytochemical study. Cancer 1984;53:1882–1901.

84 Hou-Jensen K, Priori E, Dmochowski L: Studies on ultrastructure of Ewing's sarcoma of bone. Cancer 1972;29:280–286.

85 Llombart-Bosch A, Blache R, Peydro-Olaya: Ultrastructural study of 28 cases of Ewing's sarcoma: Typical and atypical forms. Cancer 1978;41:1362–1373.

86 Nascimento AG, Unni KK, Pritchard DJ, et al: A clinicopathologic study of 20 cases of large cell (atypical) Ewing's sarcoma of bone. Am J Surg Pathol 1980;4:29–36.

87 Hartman KR, Triche TJ, Kinsella TJ, et al: Histopathology: A significant prognostic factor in Ewing's sarcoma. A review of 56 cases of distal extremity primary tumors. Cancer 1991;67:163–171.

88 Nakayama I, Tsuda N, Fuji H, et al: Fine structural comparison of Ewing's sarcoma with neuroblastoma. Acta Pathol Jpn 1975;25:251–268.

89 Enzinger FM, Weiss SW: Soft Tissue Tumors. St Louis, Mosby, 1983, pp 801–810.

90 Ushigome S, Shimoda T, Takaki K, et al: Immunocytochemical and ultrastructural studies of the histogenesis of Ewing's sarcoma and putatively related tumors. Cancer 1989;64:52–62.

91 Angervall L, Enzinger FM: Extraskeletal neoplasm resembling Ewing's sarcoma. Cancer 1975;36:240–251.

92 Bednár B: Solid dendritic cell angiosarcoma: re-interpretation of extraskeletal sarcoma resembling Ewing's sarcoma. J Pathol 1979;130:217–222.

93 Berthold F, Kracht J, Lampert F, et al: Ultrastructural, biochemical, and cell culture studies of a presumed extraskeletal Ewing's with special reference to the differential diagnosis from neuroblastoma. J Cancer Res Clin Oncol 1982;103(suppl):293–304.

94 Dickman PS, Triche TJ: Extraosseous Ewing's sarcoma versus primitive rhabdomyosarcoma: Diagnostic criteria and clinical correlation. Hum Pathol 1986;17:881–893.

95 Erlandson RA, Cordon-Cardo C: Neoplasms of complex or uncertain histogenesis; in Azar HA (ed): Pathology of Human Neoplasms. An Atlas of Diagnostic Electron Microscopy and Immunohistochemistry. New York, Raven Press, 1988, pp 533–611.

96 Gillespie JJ, Roth LM, Wills ER, et al: Extraskeletal Ewing's sarcoma. Histologic and ultrastructural observations in three cases. Am J Surg Pathol 1979;3:99–108.

97 Huntrakoon M: Extraskeletal Ewing's sarcoma. Ultrastruct Pathol 1987;11:411–419.

98 Soule EH, Newton W Jr, Moon TE, et al: Extraskeletal Ewing's sarcoma. A preliminary review of 26 cases encountered in the Intergroup Rhabdomyosarcoma study. Cancer 1978;42:259–264.

99 Mierau GW: Extraskeletal Ewing's sarcoma (peripheral neuroepithelioma). Ultrastruct Pathol 1985;9:91–98.

100 Llombart-Bosch A, Carda C, Peydro-Olaya A, et al: Soft tissue Ewing's sarcoma. Characterization in established cultures and xenografts with evidence of a neuroectodermic phenotype. Cancer 1990;66:2589–2601.

101 Shimada H, Newton WA Jr, Soule EH, et al: Pathologic features of extraosseous Ewing's sarcoma. A report from the Intergroup Rhabdomyosarcoma Study. Hum Pathol 1988;19:442–453.

102 Dias P, Parham DM, Shapiro DN, et al: Myogenic regulatory protein (MyoD1) expression in childhood solid tumors. Diagnostic utility in rhabdomyosarcoma. Am J Pathol 1990;137:1283–1291.

103 Garvin AJ, Stanley WS, Bennett DD, et al: The in vitro growth, heterotransplantation, and differentiation of a human rhabdomyosarcoma cell line. Am J Pathol 1986;125:208–217.

104 Dardick I, Ho SPE, McCaughey WTE: Soft-tissue sarcoma of undetermined histogenesis. An ultrastructural study. Arch Pathol Lab Med 1981;105:214–217.

105 Rice RW, Cabot A, Johnson AD: The application of electron microscopy to the differential diagnosis of Ewing's sarcoma and reticulum cell sarcoma of bone. Clin Orthop 1973;91:174–185.

106 Llombart-Bosch A, Peydro-Olaya A: Scanning and transmission electron microscopy of Ewing's sarcoma of bone (typical and atypical variants). Virchows Arch [A] 1983;398:329–346.

107 Dabska M, Huvos AG: Mesenchymal chondrosarcoma in the young: A clinicopathologic study of 19 patients with explanation of histogenesis. Virchows Arch [A] 1983;399:89–104.

108 Sim FH, Unni KK, Beabout JW, et al: Osteosarcoma with small cells simulating Ewing's tumor. J Bone Joint Surg 1979;61:207–215.

109 Bauer FCH, Mirra JM, Urist MR: Bone induction by Ewing's sarcoma. Transplantation into athymic nude mice. Arch Pathol Lab Med 1981;105:322–324.

110 Llombart-Bosch A, Peydro-Olaya A, Gomar F: Ultrastructure of one Ewing's sarcoma of bone with endothelial character and a comparative review of the vessels in 27 cases of typical Ewing's sarcoma. Pathol Res Pract 1980;167:71–87.

111 Pomeroy TC, Johnson RE: Prognostic factors for survival in Ewing's sarcoma. Cancer 1975;35:36–47.

112 Glaubiger DL, Makuch R, Schwarz J, et al: Determination of prognostic factors and their influence on therapeutic results in patients with Ewing's sarcoma. Cancer 1980;45:2213–2219.

113 Mendenhall CM, Marcus RB, Enneking WF, et al: The prognostic significance of soft-tissue extension in Ewing's sarcoma. Cancer 1983;51:913–917.

114 Daugaard S, Kamby C, Sunde LM, et al: Ewing's sarcoma: A retrospective study of histological and immunohistochemical factors and their relation to prognosis. Virchows Arch [A] 1989;414:243–251.

115 Stefani E, Carzoglio M, Pellegrini HD, et al: Ewing's sarcoma: value of tumor necrosis as a predictive factor. Bull Cancer (Paris) 1984;71:16–21.

116 Seemayer TA, Thelmo WL, Bolande RP, et al: Peripheral neuroectodermal tumors; in Rosenberg HS, Bolande RP (eds): Perspectives in Pediatric Pathology. Chicago, Year Book Medical Publishers, 1975, vol 2, pp 151–172.

117 Mackay B, Luna MA, Butler JJ: Adult neuroblastomas. Electron microscopic observations of nine cases. Cancer 1976;37:1334–1351.

118 Jürgens H, Bier V, Harms D, et al: Malignant peripheral neuroectodermal tumors. A retrospective analysis of 42 patients. Cancer 1988;61:349–357.

119 Shuangshoti S: Primitive neuroectodermal (neuroepithelial) tumour of soft tissue of the neck in a child: demonstration of neuronal and neuroglial differentiation. Histopathology 1986;10:651–658.

120 Shuangshotti S, Kasantikul V, Suwangoo P, et al: Malignant neoplasm of mixed mesenchymal and neuroepithelial origin (ectomesenchymoma) of thigh. J Surg Oncol 1984;27:208–213.

121 Shinoda M, Tsutsumi Y, Hata J-I, et al: Peripheral neuroepithelioma in childhood. Arch Pathol Lab Med 1988;112:1155–1158.

122 Hatchitanda Y, Tsuneyoshi M, Enjoji M, et al: Congenital neuroectodermal tumor with epithelial and glial differentiation. An ultrastructural and immunohistochemical study. Arch Pathol Lab Med 1990;114:101–105.

123 Axiotis CA, Merino M, Tsokos M, et al: Epithelioid malignant peripheral nerve sheath tumor with squamous differentiation: A light, ultrastructural, and immunohistochemical study. Surg Pathol 1990;3:301–308.

124 DiCarlo EF, Woodruff JM, Bansal M, et al: The purely epithelioid malignant peripheral nerve sheath tumor. Am J Surg Pathol 1986;10:478–490.

125 Harkin JC, Reed RJ: Tumors of the Nervous System. Atlas of Tumor Pathology (suppl). Washington, Armed Forces Institute of Pathology, 1983, pp 35s–38s.

126 Dehner LP: Whence the primitive neuroectodermal tumor? (editorial) Arch Pathol Lab Med 1990;114:16–17.

127 Willis RA: Pathology of Tumors. New York, Appleton-Century-Crofts, 1967.

128 Stout AP: Tumors of the Peripheral Nervous System. Atlas of Tumor Pathology. Washington, Armed Forces Institute of Pathology, 1949.

129 Herrera GA, Croft JL, Robert ER: Neurosecretory granule-like structures in lymphomas. Hum Pathol 1980;11:449–456.

130 Yunis EJ, Jaffe R: Anaplastic malignant round cell tumor of chest wall. A case for the panel. Ultrastruct Pathol 1982;3:387–392.

131 Nakamura Y, Becker LE, Maner K, et al: Peripheral medulloepithelioma. Acta Neuropathol 1982;57:137–142.

132 Zimmerman LE, Font RL, Anderson SR: Rhabdomyosarcomatous differentiation in malignant intraocular medulloepitheliomas. Cancer 1972;30:817–835.

133 Kirsch SB, Urich H: Medulloepithelioma: Definition of an entity. J Neuropathol Exp Neurol 1972;31:27–53.

134 Masson P (translated by Kobernick SD): Histology, Diagnosis, and Technique. Detroit, Wayne State University Press, 1970.

135 Cutler LS, Chaudhry AP, Topazian R: Melanotic neuroectodermal tumor of infancy: an ultrastructural study, literature review, and reevaluation. Cancer 1981;48:257–270.

136 Stowens D, Lin TH: Melanotic progonoma of the brain. Hum Pathol 1974;5:105–113.

137 Rickets RR, Majmudarr B: Epididymal melanotic neuroectodermal tumor of infancy. Hum Pathol 1985;16:416–420.

138 Navas Palacios JJ: Malignant melanotic neuroectodermal tumor. Light and electron microscopic study. Cancer 1980;46:529–536.

139 Young S, Gonzalez-Crussi F: Melanocytic neuroectodermal tumor of the foot. Report of a case with multicentric origin. Am J Clin Pathol 1985;84:371–378.

140 Dehner LP, Sibley RK, Saul JJ, et al: Malignant melanotic neuroectodermal tumor of infancy. A clinical, pathologic, ultrastructural and tissue culture study. Cancer 1979;43:1389–1410.

141 Stirling RW, Powell G, Fletcher CDM: Pigmented neuroectodermal tumor of infancy: An immunohistochemical study. Histopathology 1988;12:425–435.

142 Taira Y, Nalayama I, Takahara O, et al: Histological and fine structural studies on pigmented neuroectodermal tumor of infancy. Acta Pathol Jpn 1978;28:83–98.

143 Lurie HI: Congenital melanocarcinoma, melanotic adamantinoma, retinal anlage tumor, progonoma, and pigmented epulis of infancy. Summary and review of the literature and report of the first case in an adult. Cancer 1961;14:1090–1108.

144 Nikai H, Ijuhin N, Yamasaki A, et al: Ultrastructural evidence for neural crest origin of the melanotic neuroectodermal tumor of infancy. J Oral Pathol 1977;6:221–232.

145 Chaudhry AP, Haar JG, Koul A, et al: Olfactory neuroblastoma (esthesioneuroblastoma). A light and ultrastructural study of two cases. Cancer 1979;44:564–579.

146 Berger L, Luc G, Richard D: L'esthésioneuroepithéliome olfactif. Bull Assoc Fr Etude Cancer 1924;13:410–421.

147 Elkon D, Hightower SI, Lim ML, et al: Esthesioneuroblastoma. Cancer 1979;44:1087–1094.

148 Taxy JB, Bharani NK, Mills SE, et al: The spectrum of olfactory neural tumors. A light-microscopic immunohistochemical and ultrastructural analysis. Am J Surg Pathol 1986;10:687–695.

149 Ng HK, Poon WS, Poon CY, et al: Intracranial olfactory neuroblastoma mimicking carcinoma: report of two cases. Histopathology 1988;12:393–403.

150 Mills SE, Frierson HF Jr: Olfactory neuroblastoma: a clinicopathologic study of 21 cases. Am J Surg Pathol 1985;9:317–327.

151 Curtis JL, Rubinstein LJ: Pigmented olfactory neuroblastoma: a new example of melanotic neuroepithelial neoplasm. Cancer 1982;49:2136–2145.

152 Oberman HA, Rice DH: Olfactory neuroblastoma: a clinicopathologic study. Cancer 1976;38:2494–2502.

153 Olsen KD, DeSanto LW: Olfactory neuroblastoma: biologic and clinical behavior. Arch Otolaryngol 1983;109:797–802.

154 Jensen KJ, Elbrond O, Lund C: Olfactory esthesioneuroblastoma. J Laryngol Otol 1976;90:1007–1013.

155 Kadish S, Goodman M, Wang CC: Olfactory neuroblastoma. A clinical analysis of 17 cases. Cancer 1976;37:1571–1576.

156 Taxy JB, Hidvegi DF: Olfactory neuroblastoma: an ultrastructural study. Cancer 1977;39:131–138.

157 Schochet SS Jr, Peters B, O'Neal J, et al: Intracranial esthesioneuroblastoma: A light and electron microscopic study. Acta Neuropathol 1975;31:181–189.

158 Karcioglu Z, Someren A, Mathes SJ: Ectomesenchymoma: A malignant tumor of migratory neural crest (ectomesenchyme) remnants showing ganglionic, Schwannian, melanocytic, and rhabdomyoblastic differentiation. Cancer 1977;39:2486–2496.

159 Kasantikul V, Shuangshoti S, Cutchavaree A, et al: Parapharyngeal malignant ectomesenchymoma: combined malignant fibrous histiocytoma and primitive neuroectodermal tumor with neuroglial differentiation. J Laryngol Otol 1987;101:508–515.

160 Kawamoto EH, Wedner N, Agostini RM, et al: Malignant ectomesenchymoma of soft tissue. Report of two cases and review of the literature. Cancer 1987;59:1791–1802.

161 Naka A, Matsumoto S, Shirai T, et al: Ganglioneuroblastoma associated with malignant mesenchymoma. Cancer 1975;36:1050–1056.

162 Schmidt D, Mackay B, Osborne BM, et al: Recurring congenital lesion of the cheek. The quarterly case. Ultrastruct Pathol 1982;3:85–90.

163 Shuangshoti S, Cutchavaree V: Parapharyngeal neoplasm of mixed mesenchymal and neuroepithelial origin. Arch Otolaryngol 1980;106:361–364.

164 Shuangshoti S, O'Charoen S: Cerebellar neoplasm of mixed mesenchymal and neuroepithelial origin: case report. J Neurosurg 1983;69:337–343.

165 Sirikulchayanota V, Wongwaisayawan S: Malignant ectomesenchymoma (neoplasm of mixed mesenchymal and neuroepithelial origin) of wrist joint. J Med Assoc Thai 1984;67:356–361.

166 Shimada H, Aoyama C, Chiba T, et al: Prognostic subgroups for undifferentiated neuroblastoma. Immunohistochemical study with anti-S-100 protein antibody. Hum Pathol 1985;16:471–476.

167 Hamilton G, Fellinger EJ, Schratter I, et al: Characterization of a human endocrine tissue and tumor-associated Ewing's sarcoma antigen. Cancer Res 1988;48:6127–6134.

168 Lipinski M, Braham K, Philip I, et al: Neuroectoderm-associated antigens on Ewing's sarcoma cell lines. Cancer Res 1987;47:183–187.

169 Löning Th, Liebsch J, Delling G: Osteosarcomas and Ewing's sarcomas. Comparative immunocytochemical investigation of filamentous proteins and cell membrane determinants. Virchows Arch [A] 1985;407:323–336.

170 Miettinen M, Lehto V-P, Virtanen I: Histogenesis of Ewing's sarcoma. An evaluation of intermediate filaments and endothelial cell markers. Virchows Arch [B] 1982; 41:277–284.

171 Coindre J-M, De Mascarel A, Trojani M, et al: Immunohistochemical study of rhabdomyosarcoma. Unexpected staining with S100 protein and cytokeratin. J Pathol 1988;155:127–132.

172 Miettinen M, Rapola J: Immunohistochemical spectrum of rhabdomyosarcoma and rhabdomyosarcoma-like tumors. Expression of cytokeratin and the 68-kd neurofilament protein. Am J Surg Pathol 1989;13:120–132.

173 Navarro S, Noguera R, Llombart-Bosch A, et al: Cytokeratin expression in rhabdomyosarcoma (abstract). Lab Invest 1990;62:72A.

174 Miettinen M: Immunoreactivity for cytokeratin and epithelial membrane antigen in leiomyosarcoma. Arch Pathol Lab Med 1988;112:637–657.

175 Schmechel DE: γ-Subunit of the glycolytic enzyme enolase: Nonspecific or neuron specific? (editorial). Lab Invest 1985;52:239–242.

176 Marangos PJ, Zomzely-Neurath C, York C: Determination and characterization of neuron-specific protein (NSP) associated enolase activity. Biochem Biophys Res Commun 1976;68:1309–1316.

177 Schmechel D, Marangos PJ, Brightman MW: Neuron-specific enolase is a molecular marker for peripheral and central neuroendocrine cells. Nature 1978;276:834–836.

178 Tsokos M, Linnoila RI, Chandra RS, et al: Neuron-specific enolase in the diagnosis of neuroblastoma and other small, round-cell tumors in children. Hum Pathol 1984; 15:575–584.

179 Haimoto H, Takayashi Y, Koshikawa T, et al: Immunohistochemical localization of γ-enolase in normal human tissues other than nervous and neuroendocrine tissues. Lab Invest 1985;52:257–263.

180 Pahlman S, Asscher T, Nilsson K: Expression of γ-subunit of enolase, neuron-specific enolase, in human non-neuroendocrine tumors and derived cell lines. Lab Invest 1986;54:554–560.

181 Dranoff G, Bigner DD: A word of caution in the use of neuron-specific enolase expression in tumor diagnosis (editorial). Arch Pathol Lab Med 1984;108:535.

182 Seshi B, Bell CE Jr: Preparation and characterization of monoclonal antibodies to human neuron-specific enolase. Hybridoma 1985;4:13–25.

183 Seshi B, True L, Carter D, et al: Immunohistochemical characterization of a set of monoclonal antibodies to human neuron-specific enolase. Am J Pathol 1988;131: 258–269.

184 Kawaguchi K, Koike M: Neuron-specific enolase and Leu-7 immunoreactive small round cell neoplasm. The relationship to Ewing's sarcoma in bone and soft tissue. Am J Clin Pathol 1986;85:79–83.

185 Swanson PE, Lillemoe TJ, Manivel C, et al: Mesenchymal chondrosarcoma. An immunohistochemical study. Arch Pathol Lab Med 1990;114:943–948.

186 Carlei F, Polak JM, Ceccamea A, et al: Neuronal and glial markers in tumors of neuroblastic origin. Virchows Arch [A] 1984;404:313–324.

187 Hachitanda Y, Tsuneyoshi M, Enjoji M: Expression of pan-neuroendocrine proteins in 53 neuroblastic tumors. Arch Pathol Lab Med 1989;113:381–384.

188 Choi HS, Anderson PJ: Immunohistochemical diagnosis of olfactory neuroblastoma. J Neuropathol Exp Neurol 1985;44:18–31.

189 Liem RK, Yen SE, Salomon GD, et al: Intermediate filaments in nervous tissues. J Cell Biol 1978;79:637–645.

190 Zackroff RV, Idler WW, Steinert PM, et al: In vitro reconstitution of intermediate filaments from mammalian neurofilament triplet polypeptides. Proc Natl Acad Sci USA 1982;79:754–757.

191 Molenaar WM, Baker DL, Pleasure D, et al: The neuroendocrine and neural profiles of neuroblastomas, ganglioneuroblastomas, and ganglioneuromas. Am J Pathol 1990;136:375–382.

192 Osborn M, Altmannsberger M, Shaw G, et al: Various sympathetic derived human tumors differ in neurofilament expression. Virchows Arch [B] 1982;40:141–156.

193 Trojanowski JQ, Lee YM-Y: Anti-neurofilament monoclonal antibodies: Reagents for the evaluation of human neoplasms. Acta Neuropathol 1983;59:155–158.

194 Trojanowski JQ, Lee V, Pillsbury N, et al: Neuronal origin of human esthesioneuroblastoma demonstrated with anti-neurofilament monoclonal antibodies. N Engl J Med 1982;307:159–161.

195 Thiele CJ, Israel MA: Regulation of N-myc expression is a critical event controlling the ability of human neuroblasts to differentiate. Exp Cell Biol 1988;56:321–333.

196 Bennett GS: Changes in intermediate filament composition during neurogenesis. Curr Top Dev Biol 1987;21:151–183.

197 Leong AS, Gilham PN: The effects of progressive formaldehyde fixation on the preservation of tissue antigens. Pathology 1989;21:266–268.

198 Tsokos M, Jefferson J, Patterson K, et al: Neurofilament triplet proteins and β_2-microglobulin in the diagnosis of neuroblastoma and primitive neuroectodermal tumors (abstract). Lab Invest 1989;60:98A.

199 Artlieb U, Krepler R, Wiche G: Expression of microtubule-associated proteins, Map-1 and Map-2, in human neuroblastomas and differential diagnosis of immature neuroblasts. Lab Invest 1985;53:684–691.

200 Wang C, Asai DJ, Lazarides E: The 68,000-dalton neurofilament associated polypeptide is a component of nonneuronal cells and of skeletal myofibrils. Proc Natl Acad Sci USA 1980;77:1541–1545.

201 Moll R, Lee I, Gould VE, et al: Immunocytochemical analysis of Ewing's tumors. Patterns of expression of intermediate filaments and desmosomal proteins indicate cell-type heterogeneity and pluripotential differentiation. Am J Pathol 1987; 127:288–304.

202 Lawson CW, Fisher C, Gatter KC: An immunohistochemical study of differentiation in malignant fibrous histiocytoma. Histopathology 1987;11:375–383.

203 Fisher DZ, Chaudhary N, Blobel G: cDNA sequencing of nuclear lamins A and C reveals primary and secondary structural homology to intermediate filament proteins. Proc Natl Acad Sci USA 1986;83:6450–6454.

204 Gown AM, Vogel AM: Monoclonal antibodies to intermediate filament proteins of human cells: unique and cross-reacting antibodies. J Cell Biol 1982;95:414–424.

205 Olmsted JB: Microtubule-associated proteins. Annu Rev Cell Biol 1986;2:419–455.

206 Lee M, Tuttle JB, Frankfurter A: Pattern of expression of a neuron-specific β-tubulin in early developmental stages of the chick embryo (abstract). Neurosci Abstr 1987;13:701.

207 Caccamo D, Katsetos CD, Herman M, et al: Immunohistochemistry of a spontaneous murine teratoma with neuroepithelial differentiation. Neuron-associated β-tubulin as a marker for primitive neuroepithelium. Lab Invest 1989;60:390–398.

208 Weingarten MD, Lockwood AH, Hwo SY, et al: A protein factor essential for microtubule assembly. Proc Natl Acad Sci USA 1975;72:1858–1862.

209 Wiche G, Briones E, Hirt H, et al: Differential distribution of microtubule-associated proteins MAP-1 and MAP-2 in neurons of rat brain and association of MAP-1 with microtubules of neuroblastoma cells (clone N2A). EMBO J 1983;2:1915–1920.

210 Wiche G, Briones E, Koszka C, et al: Widespread occurrence of polypeptides related to neurotubule-associated proteins (MAP-1 and MAP-2) in non-neuronal cells and tissues. EMBO J 1984;3:991–998.

211 Gould VE: A new and promising pan-NE marker. Arch Pathol Lab Med 1987;11:791–794.

212 Chang T-K, Li C-Y, Smithson WA: Immunocytochemical study of small round cell tumors in routinely processed specimens. Arch Pathol Lab Med 1989;113:1343–1348.

213 Wick MR, Stanley SJ, Swanson PE: Immunohistochemical diagnosis of sinonasal melanoma, carcinoma, and neuroblastoma with monoclonal antibodies HMB-45 and anti-synaptophysin. Arch Pathol Lab Med 1988;112:616–620.

214 Hoog A, Gould VE, Grimelius L, et al: Tissue fixation methods alter the immunohistochemical demonstrability of synaptophysin. Ultrastruct Pathol 1988;12:673–678.

215 Leclerc N, Beesley PW, Brown I, et al: Synaptophysin expression during synaptogenesis in the rat cerebellar cortex. J Comp Neurol 1989;280:197–212.

216 Angelleti RH: Chromogranins and neuroendocrine secretion (editorial). Lab Invest 1986;55:387–390.

217 Ehrhart M, Grube D, Bader M-F, et al: Chromogranin in the pancreatic islet: Cellular and subcellular distribution. J Histochem Cytochem 1986;34:1673–1682.

218 O'Connor DT, Burton D, Deftos LJ: Immunoreactive human chromogranin A in diverse polypeptide hormone producing human tumors and normal endocrine tissues. J Clin Endocrinol Metab 1983;57:1084–1086.

219 Lloyd RV, Wilson BS: Detection of chromogranin in endocrine cells with secretory granules with a monoclonal antibody (abstract). Lab Invest 1984;50:35A.

220 Hokfelt T, Elfvin LG, Schultzberg M, et al: Immunohistochemical evidence of vasoactive intestinal polypeptide-containing neurons and nerve fibers in sympathetic ganglia. Neuroscience 1977;2:885–896.

221 Bryant MG, Polak JM, Modlin I, et al: Possible dual role for vasoactive intestinal peptide as gastrointestinal hormone and neurotransmitter substance. Lancet 1976;i: 991–993.

222 Pence JC, Shorter NA: In vitro differentiation of human neuroblastoma cells caused by vasoactive intestinal peptide. Cancer Res 1990;50:5177–5183.

223 Mitchell CH, Sinatra FR, Crast FW, et al: Intractable watery diarrhea, ganglioneuroblastoma and vasoactive intestinal peptide. J Pediatr 1976;89:593–595.

224 Mendelsohn G, Eggleston JC, Olson JL, et al: Vasoactive intestinal peptide and its relationship to ganglion cell differentiation in neuroblastic tumors. Lab Invest 1979;41:144–149.

225 Nakajima T, Watanabe S, Sato Y, et al: An immunoperoxidase study of S-100 protein distribution in neoplastic tissues. Am J Surg Pathol 1982;6:715–727.

226 Stefansson K, Wollmann R, Jerkovic M: S-100 protein in soft tissue tumors derived from Schwann cells and melanocytes. Am J Pathol 1982;106:261–268.

227 Weiss SW, Langloss JM, Enzinger FM: Value of S-100 protein in the diagnosis of soft tissue tumors with particular reference to benign and malignant Schwann cell tumors. Lab Invest 1983;49:299–308.

228 Drier JK, Swanson PE, Cherwitz DL, et al: S-100 immunoreactivity in poorly differentiated carcinomas. Immunohistochemical comparison with malignant melanoma. Arch Pathol Lab Med 1987;111:447–452.

229 Swanson PE, Stanley MW, Scheithauer BW, et al: Primary cutaneous leiomyosarcoma. A histological and immunohistochemical study of 9 cases, with ultrastructural correlation. J Cutan Pathol 1988;15:129–141.

230 Kahn HJ, Marks A, Thom H, et al: Role of antibody to S-100 protein in diagnostic pathology. Am J Clin Pathol 1982;79:341–347.

231 Ushigome S, Takakuwa T, Shinigawa T, et al: Ultrastructure of cartilaginous tumors and S-100 protein in the tumors: with reference to the histogenesis of chondroblastoma, chondromyxoid fibroma and mesenchymal chondrosarcoma. Acta Pathol Jpn 1984;34:1285–1300.

232 Bignami A, Eng FL, Dahl D, et al: Localization of the glial fibrillary acid protein in astrocytes by immunofluorescence. Brain Res 1972;28:351–354.

233 Roessman U, Velasco ME, Sindley SD, et al: Glial fibrillary acidic protein (GFAP) in ependymal cells during development. An immunocytochemical study. Brain Res 1980;220:13–21.

234 Velasco ME, Dahl D, Roessman U, et al: Immunohistochemical localization of glial fibrillary acidic protein in human glial neoplasms. Cancer 1980;45:484–494.

235 Dahl D, Chi NH, Miles LE, et al: Glial fibrillary acidic (GFA) protein in Schwann cells: Fact or artifact? J Histochem Cytochem 1982;30:912–918.

236 Tascos NA, Parr J, Gonatas NK: Immunocytochemical study of the glial fibrillary acidic protein in human neoplasms of the central nervous system. Hum Pathol 1982; 13:454–458.

237 Jessen KR, Mirsky R: Nonmyelin-forming Schwann cells coexpress surface proteins and intermediate filaments not found in myelin-forming cells: A study of Ran-2, A5E3 antigen and glial fibrillary acidic protein. J Neurocytol 1984;13:923–934.

238 Barber PC, Lindsay RM: Schwann cells of the olfactory nerves contain glial fibrillary acidic protein and resemble astrocytes. Neuroscience 1982;7:3077–3090.

239 Ferris CA, Schnadig VJ, Quinn FB, et al: Olfactory neuroblastoma. Cytodiagnostic features in a case with ultrastructural and immunohistochemical correlation. Acta Cytol 1988;32:381–385.

240 Moll R, Franke WW, Schiller DL, et al: The catalog of human cytokeratins: Patterns of expression in normal epithelia, tumors and cultured cells. Cell 1982;31:11–24.

241 Franke WW, Jahn L, Knapp AC: Cytokeratins and Desmosomal Proteins in Certain Epithelioid and Non-Epithelial Cells: Cytoskeletal Proteins in Tumor Diagnosis. New York, Cold Spring Harbor Laboratory, 1989, pp 151–172.

242 McNutt MA, Bolen JW, Gown AM, et al: Coexpression of intermediate filaments in human epithelial neoplasms. Ultrastruct Pathol 1985;9:31–43.

243 Azumi N, Battifora H: The distribution of vimentin and keratin in epithelial and nonepithelial neoplasms. A comprehensive immunohistochemical study on formalin- and alcohol-fixed tumors. Am J Clin Pathol 1987;88:286–296.

244 Kahn HJ, Yeger H, Baumal R, et al: Categorization of pediatric neoplasms by immunostaining with antiprekeratin and antivimentin antisera. Cancer 1983; 51:645–653.

245 Noguera R, Navarro S, Llombart-Bosch A, et al: Immunocytochemical analysis of differentiated and undifferentiated Ewing's sarcoma in vitro with a panel of antibodies (abstract). Lab Invest 1990;62:6P(31).

246 Miettinen M, Lehto V-P, Virtanen I: Histogenesis of Ewing's sarcoma. An evaluation on intermediate filaments and endothelial cell markers. Virchows Arch [B] 1982; 41:277–284.

247 Llombart-Bosch A, Carda C, Boix J, et al: Value of nude mice xenografts in the expression of cell heterogeneity of human sarcomas of bone and soft tissue. Pathol Res Pract 1988;183:683–692.

248 Knapp AC, Franke WW: Spontaneous losses of control of cytokeratin gene expression in transformed, non-epithelial human cells occurring at different levels of regulation. Cell 1989;59:67–79.

249 Cowin P, Kapprell H-P, Franke WW: The complement of desmosomal plaque proteins in different cell types. J Cell Biol 1985;101:1442–1454.

250 Schmelz M, Duden R, Cowin P, et al: A constitutive transmembrane glycoprotein of M_r 165,000 (desmoglein) in epidermal and non-epidermal desmosomes. II. Immunolocalization and microinjection studies. Eur J Cell Biol 1986;42:184–199.

251 Steinert PM, Steven AC, Roop DR: The molecular biology of intermediate filaments. Cell 1985;42:411–419.

252 Skalli O, Gabbiani G, Babai F, et al: Intermediate filament proteins and actin isoforms as markers for soft tissue tumor differentiation and origin. II. Rhabdomyosarcomas. Am J Pathol 1988;130:515–531.

253 Gard DL, Lazarides E: The synthesis and distribution of desmin and vimentin during myogenesis in vitro. Cell 1980;19:263–275.

254 Molenaar WM, Oosterhuis JW, Oosterhuis AM, et al: Mesenchymal and muscle-specific intermediate filaments (vimentin and desmin) in relation to differentiation in childhood rhabdomyosarcomas. Hum Pathol 1985;16:838–843.

255 Giorno R, Sciotto C: Use of monoclonal antibodies for analyzing the distribution of the intermediate filament protein vimentin in human non-Hodgkin's lymphomas. Am J Pathol 1985;120:351–355.

256 Tapscott SJ, Bennett GS, Toyama Y, et al: Intermediate filament proteins in the developing chick spinal cord. Dev Biol 1981;86:40–54.

257 Franke WW, Schmid E, Winter S, et al: Widespread occurrence of intermediate-sized filaments of the vimentin-type in cultured cells from diverse vertebrates. Exp Cell Res 1979;123:25–46.

258 Azumi N, Ben-Ezra J, Battifora H: Immunophenotypic diagnosis of leiomyosarcomas and rhabdomyosarcomas with monoclonal antibodies to muscle-specific actin and desmin in formalin-fixed tissue. Mod Pathol 1988;1:469–474.

259 Miettinen M, Lehto V-P, Badley RA, et al: Alveolar rhabdomyosarcoma. Demonstration of the muscle type intermediate filament protein, desmin, as a diagnostic aid. Am J Pathol 1982;108:246–251.

260 Tsokos M: The role of immunocytochemistry in the diagnosis of rhabdomyosarcoma (editorial). Arch Pathol Lab Med 1986;110:776–778.

261 Tsukada T, McNutt MA, Ross R, et al: HHF35, a muscle actin-specific monoclonal antibody. II. Reactivity in normal, reactive, and neoplastic human tissues. Am J Pathol 1987;127:389–402.

262 Burt AD, Robertson JL, Heir J, et al: Desmin-containing stellate cells in rat liver: Distribution in normal animals and response to experimental acute liver injury. J Pathol 1986;150:29–35.

263 Dahl D, Bignami A: Immunohistological localization of desmin, the muscle-type 100 A filament protein, in rat astrocytes and Muller glia. J Histochem Cytochem 1982; 30:207–213.

264 Gould VE, Rorke LB, Jansson DS, et al: Primitive neuroectodermal tumors of the central nervous system express neuroendocrine markers and may express all classes of intermediate filaments. Hum Pathol 1990;21:245–252.

265 Roholl PJM, Elbers HR, Prinsen I, et al: Distribution of actin isoforms in sarcomas. An immunohistochemical study. Hum Pathol 1990;21:1269–1274.

266 De Jong ASH, van Kessel-van Vark M, Albus-Lutter ChE, et al: Skeletal muscle actin as tumor marker in the diagnosis of rhabdomyosarcoma in childhood. Am J Surg Pathol 1985;9:467–474.

267 Tsokos M, Howard R, Costa J: Immunohistochemical study of alveolar and embryonal rhabdomyosarcoma. Lab Invest 1983;48:148–155.

268 Warnke RA, Gatter KC, Phil D, et al: Diagnosis of human lymphoma with monoclonal antileukocyte antibodies. N Engl J Med 1983;309:1275–1281.

269 Andres TL, Kadin ME: Immunologic markers in the differential diagnosis of small round cell tumors from lymphocytic lymphoma and leukemia. Am J Clin Pathol 1983;79:546–552.

270 Gonzalez-Crussi F, Mangkornkanok M, Hsueh W: Large-cell lymphoma: diagnostic difficulties and case study. Am J Surg Pathol 1987;11:59–65.

271 Gibson FM, Kemshead JT: A monoclonal antibody (FMG25) that can differentiate neuroblastoma from other small round-cell tumors of childhood. Int J Cancer 1987;39:554–559.

272 Kemshead JT, Fritschy J, Asser U, et al: Monoclonal antibodies defining markers with apparent selectivity for particular haemopoietic cell types may also detect antigens on cells of neural crest origin. Hybridoma 1982;1:109–123.

273 Sugimoto T, Sawada T, Arakawa S, et al: Possible differential diagnosis of neuroblastoma from rhabdomyosarcoma and Ewing's sarcoma by using a panel of monoclonal antibodies. Jpn J Cancer Res (Gann) 1985;76:301–307.

274 Kahn HJ, Thorner PS: Monoclonal antibody MB2: A potential marker for Ewing's sarcoma and primitive neuroectodermal tumor. Pediatr Pathol 1989;9:153–162.

275 Caillaud J-M, Benjelloun S, Bosq J, et al: HNK-1-defined antigen detected in paraffin-embedded neuroectoderm tumors and those derived from cells of the amine precursor uptake and decarboxylation system. Cancer Res 1984;44:4432–4439.

276 McGarry RC, Helfand SL, Quarles RH, et al: Recognition of myelin-associated glycoprotein by the monoclonal antibody HNK-1. Nature 1983;306:376–378.

277 Kruse J, Mailhammer R, Wernecke H, et al: Neural cell adhesion molecules and myelin-associated glycoprotein share a common carbohydrate moiety recognized by monoclonal antibodies L2 and HNK-1. Nature 1984;311:153–155.

278 Chou KH, Ilyas AA, Evans JE, et al: Structure of a glycolipid reacting with monoclonal IgM in neuropathy and with HNK-1. Biochem Biophys Res Commun 1985;128:383–388.

279 Michels S, Swanson PE, Robb JA, et al: Leu-7 in small cell neoplasms. An immuno-histochemical study with ultrastructural correlations. Cancer 1987;60:2958–2964.

280 Reynolds CP, Smith RG: A sensitive immunoassay for human neuroblastoma cells; in Mitchell MS, Oettgen HF (eds): Hybridomas in Cancer Diagnosis and Treatment. New York, Raven Press, 1980, pp 235–240.

281 Matsumura T, Sugimoto T, Sawada T, et al: Cell surface membrane antigen present on neuroblastoma cells but not fetal neuroblasts recognized by a monoclonal antibody (KP-NAC8). Cancer Res 1987;47:2924–2930.

282 Gross N, Beck D, Carrel S, et al: Highly selective recognition of human neuroblas-toma cells by mouse monoclonal antibody to a cytoplasmic antigen. Cancer Res 1986;46:2988–2994.

283 Oppedal BR, Brandtzaeg P, Kemshead JT: Immunohistochemical differentiation of neuroblastomas from other small round cell neoplasms of childhood using a panel of mono- and polyclonal antibodies. Histopathology 1987;11:363–374.

284 Patel K, Rossell RJ, Bourne S, et al: Monoclonal antibody UJ13A recognizes the neural cell adhesion molecule (NCAM). Int J Cancer 1989;44:1062–1068.

285 Jin L, Hemperly J, Lloyd RV: Expression of neural cell adhesion molecule in normal and neoplastic human neuroendocrine tissues. Am J Pathol 1991;138:961–969.

286 N'Guyen C, Mattei MG, Mattei JF, et al: Localization of the human N-CAM gene to band q23 of chromosome 11: the third gene coding for a cell interaction molecule mapped to the distal portion of the long arm of chromosome 11. J Cell Biol 1986;102:711–715.

287 Lipinski M, Hirsch M-R, Deagostini H, et al: Characterization of neural cell adhesion molecules (NCAM) expressed by Ewing and neuroblastoma cell lines. Int J Cancer 1987;40:81–86.

288 Parham P, Barnstable C, Bodmer W: Use of a monoclonal antibody (W6/32) in structural studies of HLA-A,B,C antigens. J Immunol 1979;123:342–349.

289 Hood L, Steinmetz M, Goodenow R: Genes of the major histocompatibility com-plex. Cell 1982;28:685–687.

290 Bodmer WF: HLA structure and function: a contemporary view. Tissue Antigens 1981;17:9–20.

291 Whelan JP, Chatten J, Lampson LA: HLA class I and β_2-microglobulin expression in frozen and formaldehyde-fixed paraffin sections of neuroblastoma tumors. Cancer Res 1985;45:5976–5983.

292 Lampson LA, Fisher CA, Whelan JP: Striking paucity of HLA-A,B,C and β_2-microglobulin on human neuroblastoma cell lines. J Immunol 1983;130:2471–2478.

293 Gross N, Beck D, Favre S, et al: In vitro antigenic modulation of human neuroblas-toma cells induced by IFN-γ, retinoic acid and dibutyryl cyclic AMP. Int J Cancer 1987;39:521–529.

294 Evans A, Main E, Zier K, et al: The effects of gamma interferon on the natural killer and tumor cells of children with neuroblastoma. A preliminary report. Cancer 1989;64:1383–1387.

295 Squire R, Fowler CL, Brooke SP, et al: The relationship of class I MHC antigen expression to stage IV-S disease and survival in neuroblastoma. J Pediatr Surg 1990;25:381–386.

296 Schwartz RH: T-lymphocyte recognition of antigen in association with gene products of the major histocompatibility complex. Annu Rev Immunol 1985;3:237–261.

297 Doyle A, Martin WJ, Funa K, et al: Markedly decreased expression of class I histocompatibility antigens, protein, and mRNA in human small-cell lung cancer. J Exp Med 1985;161:1135–1151.

298 Whitwell HL, Hughes HP, More M, et al: Expression of major histocompatibility antigens and leukocyte infiltration in benign and malignant human breast disease. Br J Cancer 1984;49:161–172.

299 Bernards R, Dessain SK, Weinberg RA: N-myc amplification causes down-modulation of MHC class I antigen expression in neuroblastoma. Cell 1986;47:667–674.

300 Ladisch S, Wu Z-L: Detection of a tumor-associated ganglioside in plasma of patients with neuroblastoma. Lancet 1985;i:136–138.

301 Fellinger EJ, Garin-Chesa P, Su SL, et al: Biochemical and genetic characterization of the HBA71 Ewing's sarcoma cell surface antigen. Cancer Res 1991;51:336–340.

302 Kovar H, Dworzak M, Strehl S, Schnell E, et al: Overexpression of the pseudoautosomal gene MIC2 in Ewing's sarcoma and peripheral primitive neuroectodermal tumors. Oncogene 1990;5:1067–1070.

303 Fellinger EJ, Garin-Chesa P, Triche TJ, et al: Immunohistochemical analysis of Ewing's sarcoma cell surface antigen p30/32^{MIC2}. Am J Pathol 1991;139:317–325.

304 Brodeur GM, Fong C-T: Molecular biology and genetics of human neuroblastoma. Cancer Genet Cytogenet 1989;41:153–174.

305 Vogel F: Genetics of retinoblastoma. Hum Genet 1979;52:1–54.

306 Matsunaga E: Genetics of Wilms' tumor. Hum Genet 1981;57:231–246.

307 Brodeur GM, Green AA, Hayes FA, et al: Cytogenetic features of human neuroblastomas and cell lines. Cancer Res 1981;41:4678–4686.

308 Gilbert F, Feder M, Balaban G, et al: Human neuroblastomas and abnormalities of chromosomes 1 and 17. Cancer Res 1984;44:5444–5449.

309 Fong CT, Dracopoli NC, White PS, et al: Loss of heterozygosity for chromosome 1p in human neuroblastomas: Correlation with N-myc amplification. Proc Natl Acad Sci USA 1989;86:3753–3757.

310 Atkin NB: Chromosome 1 aberrations in cancer. Cancer Genet Cytogenet 1986;21:279–285.

311 Rosen N, Reynolds CP, Thiele CJ, et al: Increased N-myc expression following progressive growth of human neuroblastoma. Cancer Res 1986;46:4139–4142.

312 Biedler JL, Spengler BA: Metaphase chromosome anomaly: Association with drug resistance and cell-specific products. Science 1976;191:185–187.

313 Kohl NE, Kanda N, Schreck RR, et al: Transposition and amplification of oncogene-related sequences in human neuroblastomas. Cell 1983;35:359–367.

314 Montgomery KT, Biedler JL, Spengler BA, et al: Specific DNA sequence amplification in human neuroblastoma cells. Proc Natl Acad Sci USA 1983;80:5724–5728.

315 Schwab M, Alitalo K, Klempnauer KH, et al: Amplified DNA with limited homology to myc cellular oncogene is shared by human neuroblastoma cell lines and a neuroblastoma tumour. Nature 1983;305:245–248.

316 Schwab M, Varmus HE, Bishop JM, et al: Chromosome localization in normal human cells and neuroblastomas of a gene related to c-myc. Nature 1984;308:288–291.

317 Brodeur GM, Seeger RC: Gene amplification in human neuroblastomas: Basic mechanisms and clinical implications. Cancer Genet Cytogenet 1986;19:101–111.

318 Emanuel BS, Balaban G, Boyd JP, et al: N-myc amplification in multiple homogeneously staining regions in two human neuroblastomas. Proc Natl Acad Sci USA 1984;82:3736–3740.

319 Brodeur GM, Seeger RC, Schwab M, et al: Amplification of N-myc in untreated human neuroblastomas correlates with advanced disease stage. Science 1984;224:1121–1124.

320 Brodeur GM, Seeger RC, Sather H, et al: Clinical implications of oncogene activation in human neuroblastomas. Cancer 1986;58:541–545.

321 Seeger RC, Brodeur GM, Sather H, et al: Association of multiple copies of the N-myc oncogenes with rapid progression of neuroblastomas. N Engl J Med 1985;313:1111–1116.

322 Cohn SL, Herst CV, Maurer HS, et al: N-myc amplification in an infant with stage IVS neuroblastoma. J Clin Oncol 1987;5:1441–1444.

323 Nakagawara A, Ikeda K, Tsuda T, et al: Amplification of N-myc oncogene in stage II and IVS neuroblastomas may be a prognostic indicator. J Pediatr Surg 1987;22: 415–418.

324 Slavc I, Ellenbogen R, Jung W-H, et al: myc Gene amplification and expression in primary human neuroblastoma. Cancer Res 1990;50:1459–1463.

325 Tsuda T, Obara M, Hiraqno H, et al: Analysis of N-myc amplification in relation to disease stage and histologic types in human neuroblastomas. Cancer 1987;60:820–826.

326 Tsuda H, Shimosato Y, Upton M, et al: Retrospective study of amplification of N-myc and c-myc genes in pediatric solid tumors and its association with prognosis and tumor differentiation. Lab Invest 1988;59:321–327.

327 Noguchi M, Hirohashi S, Tsuda H, et al: Detection of amplified N-myc gene in neuroblastoma by in situ hybridization using the single-step silver enhancement method. Mod Pathol 1988;1:428–432.

328 Brodeur GM, Hayes FA, Green AA, et al: Consistent N-myc copy number in simultaneous or consecutive neuroblastoma samples from sixty individual patients. Cancer Res 1987;47:4248–4253.

329 Christiansen H, Lampert F: Tumour karyotype discriminates between good and bad prognostic outcome in neuroblastoma. Br J Cancer 1988;57:121–126.

330 Hayashi Y, Kanda N, Inaba T, et al: Cytogenetic findings and prognosis in neuroblastoma with emphasis on marker chromosome 1. Cancer 1989;63:126–132.

331 Kaneko Y, Kanda N, Maseki N, et al: Different karyotypic patterns in early and advanced stage neuroblastomas. Cancer Res 1987;47:311–318.

332 Grady-Leopardi EF, Schwab M, Ablin AR, et al: Detection of N-myc oncogene expression in human neuroblastoma by in situ hybridization and blot analysis: Relationship to clinical outcome. Cancer Res 1986;46:3196–3199.

333 Kohl NE, Gee CE, Alt FW: Activated expression of the N-myc gene in human neuroblastomas and related tumors. Science 1984;226:1335–1337.

334 Nisen PD, Waber PG, Rich MA, et al: N-myc oncogene RNA expression in neuroblastoma. J Natl Cancer Inst 1988;80:1633–1637.

335 Schwab M, Ellison J, Busch M, et al: Enhanced expression of the human gene N-myc consequent to amplification of DNA may contribute to malignant progression of neuroblastoma. Proc Natl Acad Sci USA 1984;81:4940–4944.

336 Seeger RC, Wada R, Brodeur GM, et al: Expression of N-myc by neuroblastomas with one or multiple copies of the oncogene. Prog Clin Biol Res 1988;271:41–49.

337 Cohen PS, Seeger RC, Triche TJ, et al: Detection of N-myc gene expression in neuroblastoma tumors by in situ hybridization. Am J Pathol 1988;131:391–397.

338 Hashimoto H, Daimaru Y, Enjoji M, et al: N-myc gene product expression in neuroblastoma. J Clin Pathol 1989;42:52–55.

339 Ikegaki N, Bukovsky J, Kennett RH: Identification and characterization of the NMYC gene product in human neuroblastoma cells by monoclonal antibodies with defined specificities. Proc Natl Acad Sci USA 1986;83:5929–5933.

340 Triche TJ, Cavazzana AO, Navarro S, et al: N-myc protein expression in small round cell tumors. Prog Clin Biol Res 1988;271:91–101.

341 Yokoyama T, Tsukahara T, Nakagawa C, et al: The N-myc gene product in primary retinoblastomas. Cancer 1989;63:2134–2138.

342 Johnson BE, Ihde DC, Makuch RW, et al: myc Family oncogene amplification in tumor cell lines established from small cell lung cancer patients and its relationship to clinical status and course. J Clin Invest 1987;79:1629–1634.

343 Lee W-H, Murphee AL, Benedict WF: Expression and amplification of the N-myc gene in primary retinoblastoma. Nature 1985;309:458–460.

344 Nisen P, Zimmerman KA, Cotter SV, et al: Enhanced expression of the N-myc gene in Wilms' tumors. Cancer Res 1986;46:6217–6222.

345 Norris MD, Brian MJ, Vowels MR, et al: N-myc amplification in Wilms' tumor. Cancer Genet Cytogenet 1988;30:187–189.

346 Jacobovits A, Schwab M, Bishop JM, et al: Expression of N-myc in teratocarcinoma stem cells and mouse embryos. Nature 1985;318:188–191.

347 Garson JA, McIntyre P, Kemshead JT: N-myc amplification in malignant astrocytoma. Lancet 1985;ii:718–719.

348 Boultwood J, Wyllie FS, Williams ED, et al: N-myc expression in neoplasia of human thyroid C-cells. Cancer Res 1988;48:4073–4077.

349 Dias P, Kumar P, Marsden HB, et al: N-myc and c-myc oncogenes in childhood rhabdomyosarcoma (letter). J Natl Cancer Inst 1990;82:151.

350 Garson JA, Clayton J, McIntyre P, et al: N-myc oncogene amplification in rhabdomyosarcoma at relapse. Lancet 1986;i:1496.

351 Hayashi Y, Sugimoto T, Horii Y, et al: Characterization of an embryonal rhabdomyosarcoma cell showing amplification and overexpression of the N-myc. Int J Cancer 1990;45:705–711.

352 Mitani K, Kurosawa H, Suzuki A, et al: Amplification of N-myc in a rhabdomyosarcoma. Jpn J Cancer Res 1986;77:1062–1065.

353 Kouraklis G, Triche TJ, Tsokos M: Amplification of c-myc in human rhabdomyosarcoma: lack of association with expression, histology, or biologic behavior (abstract). FASEB J 1988;2:A806.

354 Rouah E, Wilson DR, Armstrong DL, et al: N-myc amplification and neuronal differentiation in human primitive neuroectodermal tumors of the central nervous system. Cancer Res 1989;49:1797–1801.

355 Heikkila R, Schwab G, Wickstrom E, et al: A c-myc anti-sense oligfromodeoxynuc-

leotide inhibits entry into S phase but not progress from G_0 to G_1. Nature 1987;328:445–449.

356 Sejersen TM, Rahm M, Szabo G, et al: Similarities and differences in the regulation of N-myc and c-myc genes in murine embryonal carcinoma cells. Exp Cell Res 1987; 172:304–317.

357 Hammerling U, Bjelfnab C, Pahlman S: Different regulation of N- and c-myc expression during phorbol ester-induced maturation of human SH-SY5Y neuroblastoma cells. Oncogene 1987;2:73–77.

358 Pasquale SR, Jones GR, Doersen C-J, et al: Tumorigenicity and oncogene expression in pediatric cancers. Cancer Res 1988;48:2715–2719.

359 Grady EF, Schwab M, Rosenau W: Expression of N-myc and c-src during the development of fetal human brain. Cancer Res 1987;47:2931–2936.

360 Amatruda TT, Sidell N, Ranyard J, et al: Retinoic acid treatment of human neuroblastoma cells is associated with decreased N-myc expression. Biochem Biophys Res Commun 1985;126:1189–1195.

361 Thiele CJ, Reynolds CP, Israel MA: Decreased expression of N-myc precedes retinoic acid-induced morphological differentiation of human neuroblastoma. Nature 1985;313:404–406.

362 Thaller C, Eichelle G: Identification and spatial distribution of retinoids in the developing chick limb bud. Nature 1987;327:625–628.

363 Akeson R, Bernards R: N-myc down-regulates neural cell adhesion molecule expression in rat neuroblastoma. Mol Cell Biol 1990;10:2012–2016.

364 Breit S, Schwab M: Suppression of MYC by high expression of NMYC in human neuroblastoma cells. J Neurosci Res 1989;24:21–28.

365 Shimizu K, Goldfarb M, Perucho M, et al: Isolation and preliminary characterization of the transforming gene of a human neuroblastoma cell line. Proc Natl Acad Sci USA 1983;80:383–387.

366 Taparowsky E, Shimizu K, Goldfarb M, et al: Structure and activation of the human N-ras gene. Cell 1983;34:581–586.

367 Ballas K, Lyons J, Janssen JWG, et al: Incidence of ras gene mutations in neuroblastoma. Eur J Pediatr 1988;147:313–314.

368 Ireland CM: Activated N-ras oncogenes in human neuroblastoma. Cancer Res 1989; 49:5530–5533.

369 Tanaka T, Slamon DJ, Shimoda H, et al: Expression of Ha-ras oncogene products in human neuroblastomas and the significant correlation with patient's prognosis. Cancer Res 1988;48:1030–1034.

370 Bolen JB, Rosen N, Israel MA: Increased pp60[s-src] tyrosyl kinase activity in human neuroblastomas is associated with amino-terminal tyrosine phosphorylation of the src gene product. Proc Natl Acad Sci USA 1985;82:7275–7279.

371 Horii Y, Sugimoto T, Sawada TY, et al: Differential expression of N-myc and c-src proto-oncogenes during neuronal and Schwannian differentiation of human neuroblastoma cells. Int J Cancer 1989;43:305–309.

372 Curt GA, Carney DN, Cowan KH, et al: Unstable methotrexate resistance in human small-cell carcinoma associated with double minute chromosomes. N Engl J Med 1983;308:199–202.

373 Bates SE, Mickley LA, Chen YN, et al: Expression of a drug resistance gene in human neuroblastoma cell lines: modulation by retinoic acid-induced differentiation. Mol Cell Biol 1989;9:4337–4344.

374 Bates SE, Shieh C-Y, Tsokos M: Expression of mdr-1/P-glycoprotein in human neuroblastoma. Am J Pathol 1991;139:305–315.

375 Nakagawara A, Kadomatsou K, Sato S-I, et al: Inverse correlation between expression of multidrug resistance gene and N-myc oncogene in human neuroblastomas. Cancer Res 1990;50:3043–3047.

376 Bourthis J, Benard J, Hartman O, et al: Correlation of MDR1 gene expression with chemotherapy in neuroblastoma. J Natl Cancer Inst 1989;81:1401–1405.

377 Turc-Carel C, Aurias A, Mugneret F, et al: Chromosomes in Ewing's sarcoma. I. An evaluation of 85 cases and remarkable consistency of t(11;22)(q24;q12). Cancer Genet Cytogenet 1988;32:229–238.

378 Fujii Y, Hongo T, Nakagawa Y, et al: Cell culture of small round cell tumor originating in the thoracopulmonary region. Cancer 1989;64:43–51.

379 Davison EV, Pearson ADJ, Emslie J, et al: Chromosome 22 abnormalities in Ewing's sarcoma. J Clin Pathol 1989;42:797–799.

380 Grififin CA, McKeon C, Israel MA, et al: Comparison of constitutional and tumor-associated 11;22 translocations: Nonidentical breakpoints on chromosomes 11 and 22. Proc Natl Acad Sci USA 1986;83:6122–6126.

381 Whang-Peng J, Freter CE, Knutsen T, et al: t(11;22) translocation in esthesioneuroblastoma. Cancer Genet Cytogenet 1987;29:155–157.

382 Vigfusson NV, Allen LJ, Phillips JH, et al: A neuroendocrine tumor of the small intestine with a karyotype of 46,XY,t(11;22). Cancer Genet Cytogenet 1986;22:211–218.

383 Bartram CR, de Klein A, Hagemeijer A, et al: Localization of the human c-sis oncogene in Ph[1]-positive and Ph[1]-negative chronic myelocytic leukemia by in situ hybridization. Blood 1984;63:223–225.

384 Van Kessel ADG, Turc-Carel C, Deklein A, et al: Translocation of oncogene c-sis from chromosome 22 to chromosome 11 in Ewing's sarcoma-derived cell line. Mol Cell Biol 1985;5:427–429.

385 Bechet JM, Borkmann G, Lenoir GN: The c-sis oncogene is not activated in Ewing's sarcoma. N Engl J Med 1984;310:393.

386 Israel MA, Thiele CJ, Whang-Peng J, et al: Peripheral neuroepithelioma: genetic analysis of tumor derived cells lines. Prog Clin Biol Res 1985;175:161–170.

387 Rosen N, Bolen JB, Schwartz AM, et al: Analysis of pp60[c-src] protein kinase activity in human tumor cell lines and tissues. J Biol Chem 1986;261:13754–13759.

388 Vechio G, Cavazzana AO, Triche TJ, et al: Expression of the dbl proto-oncogene in Ewing's sarcomas. Oncogene 1989;4:897–900.

389 Pfeifer AMA, Kasid U, Tsokos M, et al: Implication of the c-raf-1 Proto-Oncogene in Neoplastic Transformation in vivo and in vitro: Cancer Cells/Molecular Diagnostics of Human Cancer. New York, Cold Spring Harbor Laboratory, 1989, pp 177–181.

390 Molenaar WM, Dam-Meiring A, Kamps WA, et al: DNA-aneuploidy in rhabdomyosarcomas as compared with other sarcomas of childhood and adolescence. Hum Pathol 1988;5:573–579.

391 Gansler T, Chatten J, Varello M, et al: Flow cytometric DNA analysis of neuroblastoma. Correlation with histology and clinical outcome. Cancer 1986;58:2453–2458.

392 Look AT, Hayes FA, Nitschke R, et al: Cellular DNA content as a predictor of response to chemotherapy in infants with unresectable neuroblastoma. N Engl J Med 1984;311:231–235.

393 Oppedal BR, Storm-Mathisen I, Lie SO, et al: Prognostic factors in neuroblastoma: Clinical, histopathologic, and immunohistochemical features and DNA ploidy in relation to prognosis. Cancer 1988;62:772–780.

394 Taylor SR, Blatt J, Costantino JP, et al: Flow cytometric analysis of neuroblastoma and ganglioneuroma: A 10-year retrospective study. Cancer 1988;62:749–754.

395 Cohn SL, Rademaker AW, Salwen HR, et al: Analysis of DNA ploidy and proliferative activity in relation to histology and N-myc amplification in neuroblastoma. Am J Pathol 1990;136:1043–1052.

396 Kaneko Y, Maseki N, Sakurai M, et al: Chromosomes and screening for neuroblastoma. Lancet 1988;i:174–175.

397 Abramowsky CR, Taylor SR, Anton AH, et al: Flow cytometry DNA ploidy analysis and catecholamine secretion profiles in neuroblastoma. Cancer 1989;63:1752–1756.

Maria Tsokos, MD, Laboratory of Pathology, National Institutes of Health, Bethesda, MD 20814 (USA)

Garvin AJ, O'Leary TJ, Bernstein J, Rosenberg HS (eds): Pediatric Molecular Pathology: Quantitation and Applications. Perspect Pediatr Pathol. Basel, Karger, 1992, vol 16, pp 99–119

Application of the Polymerase Chain Reaction to Archival Material

Beverly B. Rogers

Department of Pathology, Women and Infant's Hospital and
Brown University Medical School, Providence, R.I., USA

History

The mountains of California, during a late night drive, provided the location for the discovery of the polymerase chain reaction (PCR), one of the most powerful tools of molecular biology. Dr. Kary Mullis, at that time a member of Cetus Corporation, recalls mentally designing the PCR as he was driving a 3-hour stretch of road to his cabin [1]. The following Monday, when he returned to work at Cetus, he described his idea to 'anyone who would listen'. While there was no reason for its not working, no one remembered having seen it tried. Over the next few months, Dr. Mullis perfected the conditions for the PCR and, in 1987, published the definitive procedure [2]. Even before this, in December 1985, application of the procedure in a clinical setting had been reported for the diagnosis of sickle cell anemia [3].

In the mid-1980s, the PCR procedure required the investigator to be stationed in front of three water baths, changing reaction tubes from one to another every 2–5 min, adding a new aliquot of enzyme approximately every 10 min, and keeping track of the number of cycles. Luckily, modifications in the enzyme permitted automation of PCR [4].

Technique

The PCR uses four reagents to increase target DNA up to 1 million-fold. The reagents are: (1) DNA polymerase; (2) primers; (3) free nucleotides, and (4) template.

The *DNA polymerase* enzyme, found in eukaryotes and prokaryotes, can be isolated from bacteria or produced from the cloned gene. In vivo the

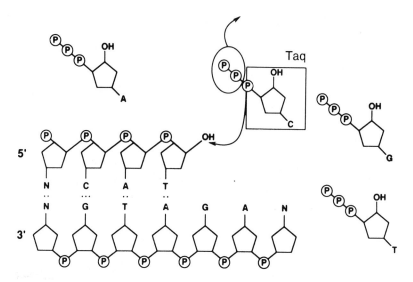

Fig. 1. The Taq polymerase uses free nucleotides placed into the PCR mixture to synthesize DNA. The polymerase, shown schematically in the box and labeled Taq, is adding a cytosine triphosphate to the growing chain. The cytosine (C) will bind with guanine (G) liberating two phosphate groups during the process. The next nucleotide required will be thymidine triphosphate (T), which will result in a hydrogen bond with adenine (A). In this way, the DNA will be synthesized in a conservative manner, preserving fidelity of the nucleotide chain. The three dots represent three hydrogen bonds between GC pairs, and two dots represent two hydrogen bonds between AT pairs. [Reprinted with permission, Rhode Island Med J 1991;74:13–16.]

enzyme is active in cell replication and repair of damaged cellular DNA and in vitro is used in the PCR to synthesize DNA. The polymerase recognizes DNA in partially single-stranded form and synthesizes a new strand of DNA from the single-stranded template (fig. 1). Single-stranded DNA is formed from double-stranded DNA by denaturation with heat. The polymerase is added along with short, synthetic pieces of DNA homologous to the area of interest. The synthetic pieces of DNA, called *primers,* attach to the *template* DNA strand after it is denatured (fig. 2) and 'prime' the reaction. The primer gives the polymerase a starting place from which to synthesize the complementary strand. After the primers anneal, the DNA polymerase uses the four *free nucleotides,* guanine, cytosine, thymine, and adenine, which are placed into the solution, to synthesize (extend) a new strand of DNA.

Each sequence of denaturation, annealing of primers and extension of the DNA strand by polymerase, forms a cycle requiring a change in tempera-

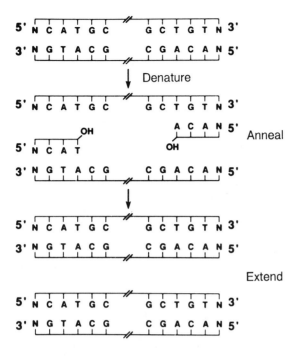

Fig. 2. The first step in a PCR cycle denatures the double-stranded target DNA, also known as the template, shown at the top of the figure. The primers, with hydroxyl groups attached, anneal to the denatured template as the temperature of the reaction is lowered (labeled Anneal). The final step in the process is extension of the DNA by the Taq polymerase (see fig. 1). This last step in a PCR cycle is labeled Extend, and shows doubling of the DNA (two double strands of DNA at the last step vs. one double strand at the first step). The DNA doubles with each cycle, and the amount of target PCR increases exponentially as the number of cycles increases. [Reprinted with permission, Rhode Island Med J 1991;74:13–16.]

ture of the reaction mixture to optimize each step. Denaturing the DNA requires a temperature of 94 °C. Because double-stranded DNA is held together by hydrogen bonds between the individual bases adenine, guanine, cytosine, and thymine, the temperature must be high enough to break these bonds to form single-stranded from double-stranded DNA.

The temperature of the reaction mixture must be lowered for the hydrogen bonds to reform or anneal between the DNA template and the primers. Primers used for diagnostic purposes would typically be 20–25 nucleotides long, with 50% guanine-cytosine content. The temperature at which 50% of the primers will form hybrids with the specific DNA template is called the Td (temperature of dissociation). The Td can be estimated by the equation Td $(°C) = 2(A+T)+4(G+C)$, where A, T, G, and C are the four nucleotides adenine, thymine, guanine, and cytosine, respectively. The fact that 4 °C is given for each G or C, and only 2 °C for each A or T reflects the stronger bonds between GC pairs than between AT pairs. The longer the primers, the higher is the annealing temperature. Typically, the annealing temperature is slightly below the Td, which allows for a high degree of binding of the primers with the specific template, but a low probability of binding of the primers with nonspecific template.

The initial PCR experiments used the Klenow polymerase, isolated from *Escherichia coli,* with optimum activity at 37 °C. With the denaturation step at 94 °C, it was necessary to add enzyme with each PCR cycle. The discovery of the Taq polymerase permitted automation of the cycling of the PCR. Since the Taq polymerase, named after the bacterium from which it was isolated *(Thermus aquaticus),* works best at high temperatures, the extension temperature is usually 72 °C.

Analysis of PCR Product

There are many methods for evaluating amplification products. Direct gel analysis is the fastest but the least sensitive means of confirming amplification. After performing the PCR, an aliquot of the reaction mixture (frequently 10–15 μl) is applied to a gel for electrophoresis. Ethidium bromide, a fluorescent chemical that binds to DNA, is added to the solution during or after electrophoresis, and the DNA is visualized as a brightly fluorescing band under ultraviolet light. Direct gel analysis is adequate in specimens containing a large amount of purified template, e.g. amplification of a human gene from DNA extracted from peripheral blood lymphocytes. Direct gel analysis may also be satisfactory, but with reduced efficiency, when amplifying DNA extracted from paraffin-embedded material or crude mixtures of cells in tissue (see Effects of Fixative, below). The use of direct gel analysis alone for identification of PCR products must be evaluated for each case.

A visible band in a direct gel, which appears to be specific for the amplified product, must be confirmed by either hybridization or restriction enzyme cleavage. Confirmation by restriction enzyme cleavage is the quickest and easiest method. After cleaving the amplified DNA by the restriction enzyme, the resulting smaller products are subjected to electrophoresis. If cleavage occurs at the target site, the specific bands of the product (fig. 3) confirms the specificity of the amplified product.

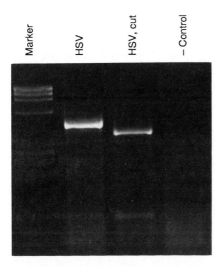

Fig. 3. Direct gel analysis of herpes simplex virus DNA amplified with the PCR. The lane marked 'HSV' shows the uncut PCR product, which measures 476 nucleotides (nt). The lane marked 'HSV, cut' is the same PCR product after cleavage with AvaII restriction enzyme, which cuts the 476 nt amplified product into bands of 389 (top band) and 87 (bottom band) nt. The negative control is an amplification using primers only; no template DNA. The molecular weight marker is a HaeIII digest of PhiX174.

Hybridization increases the sensitivity for detecting amplified DNA, but it also increases the analysis time. Hybridization is necessary when analyzing a specimen for DNA in low copy number, such as for viral DNA amidst a background of human DNA. In these cases, only hybridization and not direct gel analysis may detect amplified DNA. Hybridization may also confirm the specificity of an amplified fragment seen by direct gel analysis to the DNA of interest. The probe for hybridization is chosen in the internal portion of the amplified fragment so that revealing a band of the appropriate size confirms the amplification product.

Amplification products may also be detected by incorporating a radioactive nucleotide into the PCR reaction and separating the fragments using an acrylamide gel. The products can be autoradiographed directly, without a Southern blot. While this method increases the sensitivity over direct gel analysis, it does not confirm the specificity of the product.

Nonradioactive detection methods using biotin or digoxigenin (Boehringer Mannheim Corp., Indianapolis, Ind.) labeled probes may be used instead of a radioactive probe. Incorporation of nucleotides labeled with biotin or digoxigenin into the PCR, with transfer to a membrane by Southern blot, decreases the time to a result as compared with hybridization. Instead of autoradiography, which is necessary when using ^{32}P, the product can be visualized directly after adding the avidin conjugate (for biotin) or the antidigoxigenin antibody (for digoxigenin).

How to Design a PCR

Primers should be close enough together to allow the Taq polymerase to efficiently extend the DNA, but far enough apart so that the specificity of the amplified product can be confirmed by restriction enzyme cleavage. In general, this means primers should be approximately 100–500 nucleotides apart. The shorter the fragment amplified, the greater the chance of achieving adequate amplification from archival material since there is unavoidable DNA degradation during tissue fixation. However, shorter fragments also predispose to contamination.

In choosing primers, an important feature to guard against is a large amount of homology with each other. If the 5′ end of one primer has homology with the 3′ end of the other primer, they will tend to prime each other rather than the target DNA template, thereby decreasing the sensitivity of the PCR. The products of this primer-primer interaction are seen by direct DNA analysis using ethidium bromide-stained gels. When small, the bands are frequently referred to as 'primer-dimers', but they can also be quite large (several kilobases). Computer programs can help choose primers, but the only way to make sure a primer works well is to try it.

It is also important to avoid homology between the primers and areas of nonspecific annealing, e.g. one part of the cytomegalovirus (CMV) genome encodes a protein extensively homologous with the human class I major histocompatibility molecules [5]. Primers for amplification of CMV should be distant from this area because of the potential for cross-reactivity. Cross-reactivity of the primers with DNA other than the intended target decreases the efficiency of the amplification procedure and, with sufficient homology, an internal probe could theoretically hybridize to a nonspecifically amplified fragment giving false-positive results. Nonspecific amplification should be suspected when there are a large number of unexpected positive cases. A wide range of negative controls helps avoid this pitfall.

What concentration of primers is best for a particular PCR? Typically, 1 μM concentration (approximately 20 pmol) is used in PCR reactions, depending on the particular PCR conditions. As an example, one set of primers currently in use in our laboratory has significant complementarity of the primers with each other, but they also amplify the template DNA. An 8-fold increase in this primer set is used to increase amplification of the target DNA. To establish optimal conditions for a specific PCR reaction, it is best to prepare serial dilutions of primers.

The concentration of magnesium in the buffer is critical to the efficiency of the Taq polymerase. When establishing a PCR, varying dilutions of magnesium are used to determine the optimal concentration. In general, 1.5 mM concentration of magnesium is a workable concentration.

Most PCR protocols suggest 25–35 cycles. After 35 cycles, the amount of product is markedly decreased by the large amounts of template in relation to the reagents. The time required for denaturation, annealing, and extension varies for each PCR protocol. Usually, 1.5 min at 94 °C is sufficient for denaturation. The annealing step takes 1–2 min, and the extension step is 1–2 min for the first cycle. Adding additional time to subsequent extension steps may be advantageous as the amount of template increases. When designing a PCR, longer cycle times may be used at first, and, if amplification is efficient, the times may be decreased to result in optimal amplification in the least amount of time. The cycle times listed above may be used with the DNA Thermal Cycler (Perkin-Elmer Cetus, Norwalk, Conn.).

The annealing temperature of the primers should be high enough to prevent nonspecific annealing, but low enough to afford efficient annealing to the template. Begin by setting the annealing temperature at Td − 5 °C and, if optimal amplification does not occur, reduce the temperature. For longer primers with annealing temperatures at 65 °C or above, annealing and subsequent extension may be possible at the same temperature, resulting in shorter cycle times.

Refinements in automation (GeneAmp PCR System 9600; Perkin-Elmer Cetus) combine annealing and extension steps, decrease the cycle time, and decrease the amount of reagents used. With these newer machines, three steps in the cycle will be replaced by two steps.

Pitfalls in PCR Interpretation

A difficulty in interpreting a direct gel analysis of PCR products is to ensure that the bands visualized are specific for the PCR product, e.g. many

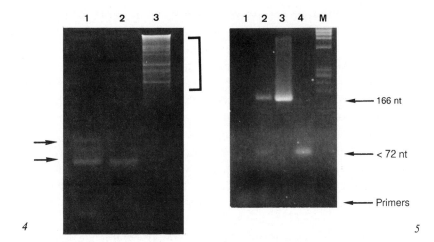

Fig. 4. Nonspecific amplification products may produce problems with interpretation. Low molecular weight nonspecific primer bands (arrows) are shown in lanes 1 and 2.
Lane 3 shows nonspecific bands from amplification of human DNA which serves as a
negative control for parvovirus B19 amplification. The bands are high molecular weight
and result from the primers binding nonspecifically to the human genome. The annealing
temperature with the primers used in lane 3 was 37 °C, below that used in most PCRs.

Fig. 5. Inhibition to amplification of an unpurified 5-μm section of formalin-fixed,
paraffin-embedded umbilical cord is demonstrated using primers to the β-globin region,
which span a 166-nt fragment. Tissue preparation consisted of deparaffinization and
hydration. The amplification products from extracted human DNA, which served as a
control, are shown in lane 3. Lane 1 is an ineffective attempt to amplify the umbilical cord
sample. Examination of the paraffin block showed a large blood clot in the umbilical vein.
In another section of the umbilical cord, the clot was physically removed, and the
amplification was repeated with success (lane 2). One indication that amplification was
inhibited in the sample from lane 1 is the absence of the band located at less than 72 nt.
This band is a primer-dimer, whose total absence in lane 1 indicates an inhibitor was
present. Note the strong intensity of the primer-dimer band in lane 4, the negative control
without template DNA. There is usually an inverse relationship between the intensity of
the specific amplification product and the primer-dimer products.

nonspecific bands may be produced because of primer-dimer formation or
nonspecific annealing of the primers to background DNA (fig. 4). If the
specific amplified fragment is close in size to the nonspecific primer bands,
interpretation requires confirmation of the specific band.

Analysis of DNA in archival material is potentially treacherous due to
effects of fixation from both the type of fixative and the time interval between
fixation and processing. To control for the effects of fixation, one can run a set

of 'control' primers with each specimen. For example, a negative result in analysis of a paraffin-embedded sample of bone marrow for parvovirus may be due to absence of virus, degradation of the DNA by fixation, or to inhibitors in the tissue. If the DNA is of adequate quality to be amplified, a second PCR on the sample seeking a gene present in every human cell, e.g. the gene for the β-globin region which is present in two copies per cell, would be strongly positive.

The use of control primers is also indicated when inhibitors [6, 7] to the PCR reaction are suspected. Hemoglobin, a common inhibitor, or other unknown inhibitors may be present in incompletely purified samples (fig. 5).

Contamination is the most difficult problem with the PCR. To detect contamination, each PCR run must include a negative control. This may be a tube with no template DNA or a tube with DNA that is negative for the template in question. One might include a negative control for any specimen processing prior to the PCR. A variety of techniques to decrease contamination are recommended [8, 9] (Appendix A).

Effects of Fixation of Archival Material

The type and duration of fixation are the two variables most likely to have deleterious effects on the amplification of DNA by the PCR in archival material. Greer et al. [10] examined fixatives and their effects on DNA during a 24-hour period. Amplification efficiency decreased in specimens fixed for 24 h in 10% buffered formalin when the amplification product was 989 bases or longer, but not when the amplified product was 536 bases or shorter. No difference was noted in the size of extracted DNA at 1, 4, and 24 h. The average size of extracted DNA fragments for all time periods was 2,000–5,000 bases.

Using primers that span a shorter fragment (135 bases) and analyzing tissue fixed in formalin for up to 4 weeks, we found amplification slightly decreased after 72 h of fixation in 10% buffered formalin, and markedly decreased after 1 week [11] (fig. 6). DNA continues to degrade with exposure to formalin, and fixation in formalin for 1 week or longer may result in specimens unsuitable for the PCR (fig. 7). Primers which amplify a short DNA fragment, such as 150 nucleotides, have a greater likelihood of success when analyzing fixed and partially degraded DNA.

Of the several fixatives studied [10], 10% buffered neutral formalin and acetone showed the least deleterious effect on amplification after 24 h.

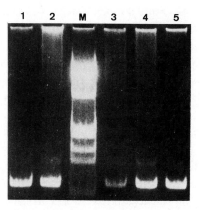

Fig. 6. Amplification products from the c-K-ras oncogene show the effects of using formalin-fixed tissue as template. Lanes 1 and 2 are varying amounts of template. Amplification after the tissue was retained 1 week in formalin (lane 3) shows a decrease in amplification compared with the intensity of the amplified products after 24 h in formalin (lane 4) or 0 h in formalin (lane 5). Molecular weight markers (HaeIII digest of PhiX 174) are designated M. [Reprinted with permission, Am J Pathol 1990;136:541–548.]

Fixatives which showed a decrease in amplification after 24 h, but which were still usually adequate for amplification included alcoholic formalin, methacarn, Zamboni's, paraformaldehyde, Clarke's, and formalin-alcohol-acetic acid. Carnoy's, Zenker's, and Bouin's were least efficient. Amplification of DNA from tissue fixed in Bouin's was possible only at 1 h with primers that spanned 110 bases. Longer fragments were not amplified and even the short 110 base pair fragment was not amplified after 4 h of fixation. Any question as to the suitability of fixed DNA for PCR amplification requires using control primers to determine if the DNA can be amplified.

The effects of specimen treatment with either proteinase K or a full DNA extraction prior to the PCR were studied, and there was little advantage of DNA extraction compared to a 3-hour proteinase K digestion when amplifying the human β-globin region [10]. This is in contrast to larger amounts of DNA recovered following a 5-day proteinase K digestion [12].

Alternatively, tissue not subjected to proteinase K treatment or purified prior to analysis [13] demonstrated variability in DNA amplification when fixed in Omnifix, formalin, ethanol, and Zenker's. The most consistently positive results were achieved using Omnifix and ethanol. Amplification of DNA was not possible when tissues were fixed in Bouin's and B-5.

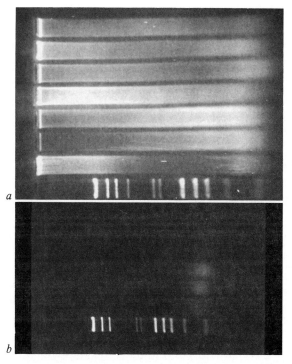

Fig. 7. Ethidium bromide-stained direct gel analysis of DNA extracted from tissue after 1–2 days in formalin *(a)* and months in formalin *(b)*. The gel is oriented with high molecular weight DNA to the left and low molecular weight DNA to the right. The DNA is examined without amplification to determine the length of the DNA. After formalin fixation for 1–2 days, the DNA appears as a smear, indicating degradation occurred. However, there is still high molecular weight DNA present above the molecular weight markers (largest marker is 23.0 kb, seen on the left of photo). However, low molecular weight products (approximately 400–500 nt in length) are the largest fragments of DNA present after months in fixative.

Amplification of RNA has also been accomplished [12–14]. Because of a large number of RNAses in cellular material, RNA is apt to degrade. Measles virus RNA from patients with SSPE was amplified using RNA extracted from paraffin-embedded tissue but not from a crude suspension of the same material [14]. Extraction of the RNA was necessary to obtain adequate results. PCR for RNA requires using a reverse transcriptase to transcribe the RNA into cDNA. After that, the PCR is identical to that for DNA.

Methods of Specimen Processing

Depending on the sensitivity necessary for a given PCR, the tissue samples may be processed in several ways. The easiest method is to place sections from the paraffin blocks into microcentrifuge tubes, deparaffinize the samples, and do the PCR directly [15] (Appendix B). This simple method works well in many cases. The drawback to this method is the potential for false-negative results due either to DNA bound by proteins or from inhibitors in the solution. In general, this procedure should be reserved for cases in which false-negative results would always be apparent, such as amplification of a single copy gene from the human genome in a sample with known human DNA, e.g. amplification of the β-globin region across the site of the sickle cell mutation, or amplification across the site of the mutation involved in cystic fibrosis. Absence of amplified fragments in either of these specimens containing human DNA requires further processing for that sample.

The most elaborate method of processing, extraction of the DNA from the tissue block, is more time consuming than the short procedure described above, but assures the greatest sensitivity in all systems (Appendix C). Extraction of DNA is used in any case to identify an infectious agent. The major drawback to performing complete extractions is the risk of contamination from one specimen to another.

Several protocols attempt to strike a balance between the simple and elaborate methods. Some involve digesting the DNA with proteinase K (with or without a detergent), as in the extraction procedure, but then boiling the DNA mixture and using the entire sample or an aliqot in the PCR [16]. Another procedure involves boiling the DNA solution followed by filtering through a Sephadex G-50 minicolumn for additional purification [17].

Application of the PCR in Analysis of Tissues

The material collected by pathologists over the years is now available for DNA analysis. The PCR provides an effective new diagnostic tool to define mutations and to identify infectious agents in archival material. Viral genome has been detected from tissue, and the PCR is particularly useful when a histologic suspicion cannot be confirmed by other methods.

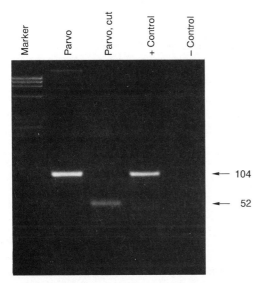

Fig. 8. Amplification of 1 μg of DNA extracted from formalin-fixed, paraffin-embedded placenta from a case of intrauterine parvovirus infection shows a bright band of amplified DNA by direct gel analysis which is 104 nt long. The amplified band from the placenta (labeled Parvo) is identical in size to the band produced by the positive control. Restriction enzyme cleavage by RsaI cuts the 104-nt-specific fragment into two 52-nt fragments, seen on the gel as a single smaller band (Parvo, cut).

Viral Detection

Parvovirus B19 causes Fifth disease in children, and may result in fetal anemia and hydrops. Characteristic, although not pathognomonic, are intranuclear inclusions in erythroid precursors. The diagnosis can be confirmed by serology, electron microscopy, or PCR analysis of the paraffin-embedded tissue. The results of DNA amplification confirmed the diagnosis in one such case in which the fetus died in utero at 20 weeks gestation with severe hydrops. The liver contained characteristic pale-to-lavender intranuclear erythroid inclusions. The placenta had many nucleated red blood cells, but no inclusions. A bright band of amplified DNA, demonstrated after PCR analysis of DNA extracted from the paraffin block of the placenta, was parvovirus DNA, as confirmed by restriction enzyme digest (fig. 8). This example underscores the use of the PCR to detect viral genome in autopsy material, but also clearly shows that inclusions for parvovirus may not be present in all infected tissues, such as the placenta in this case. Parvovirus DNA has been demonstrated in fetal liver, heart, lung, brain, and thymus both by in situ hybridization and PCR [18]. In comparing the sensitivity of the PCR with that of dot-blot hybridization in the detection of viral DNA, PCR showed a 30-fold increase in sensitivity [19].

Human immunodeficiency virus (HIV) has been detected by the PCR in paraffin-embedded lymph nodes from seropositive individuals [20]. Of 25 biopsies from seropositive patients, 23 were positive for HIV DNA by the PCR, and 19 of 20 biopsies from seronegative or low-risk individuals were negative.

CMV has been detected in DNA from paraffin-embedded blocks. In one study, PCR detected CMV genome in 4 of 10 paraffin-embedded tissue sections from 9 patients suspected of having CMV or other herpesvirus [21]. Three of the 4 tissues had inclusions characteristic of CMV and the fourth was from a patient who was culture-positive for CMV, but without inclusions. Five cases were negative for CMV histologically and by PCR and an additional case was equivocal histologically, negative for CMV using cell culture, and negative by the PCR.

PCR has been used to detect genome of JC virus in 20 of 24 patients with progressive multifocal leukoencephalopathy [22]. Two sets of primers recognized different areas of the JC virus genome, with one primer set more frequently positive than the other. Tissues from biopsies were more easily amplified than from autopsies. The 4 PCR negative cases were all from autopsy material and, in each case, amplification with 'control' primers was not successful. In situ hybridization was positive in 3 of the 4 PCR-negative cases, indicating DNA sufficiently intact to be detected by nucleic acid hybridization. An inhibitor was suspected to have interfered with amplification because the DNA was not purified prior to PCR amplification.

Measles virus genome (RNA) has been detected in both frozen tissue and formalin-fixed, paraffin-embedded tissue from the central nervous system in patients with subacute sclerosing panencephalitis [14]. Amplification was possible from specimens that had been stored frozen for up to 27 years, and stored in paraffin for 9 years. Attempts were not made to amplify DNA from samples retained for longer periods.

Other viruses that have been detected in archival material by PCR include hepatitis B virus [17], Epstein-Barr virus [23], and human papilloma virus [24].

Neoplasms and the PCR

The diagnosis of neoplasia can be supported by analysis of archival material using PCR. Monoclonality of B cells from B cell lymphomas has been confirmed by amplification across areas of the V-D-J region [25]. A clonal population gives a sharp band by direct gel analysis, whereas reactive lymph nodes produce a broad smear of amplified products; specimens from T cell lymphomas show no amplification. In cases with a suspected clonal proliferation of B cells, as in Burkitt's lymphoma, the PCR can confirm clonality. Although 2 of 26 cases of reported B cell lymphoma failed to show the single band of amplification indicative of clonality, it was not clear if a control for amplification had been used in these 2 cases.

Oncogene amplification can be detected by PCR with two sets of primers in one reaction – one set that amplifies the oncogene and another set that amplifies a 'control' gene [26] with subsequent comparison of amplified band intensity. This could be used to analyze archival material for amplification of the c-myc oncogene in neuroblastoma.

Loss of heterozygosity has been detected in tumors by microdissection to separate tumor cells from normal cells followed by PCR to amplify polymorphic sites across an area of mutation [27]. A double band pattern indicates heterozygosity; a single band, homozygosity. This technique may be applied to retinoblastomas to differentiate those cases arising in association with a constitutional mutation and those occurring sporadically.

Point mutations related to the development of tumors can be detected using the PCR to amplify DNA flanking the mutation. Hybridization using probes containing either the mutated sequence or normal sequence separates the normal from abnormal gene. Mutation in the retinoblastoma gene [28] and the K-ras oncogene [29] may be detected in this way.

The Philadelphia chromosome, seen in chronic myelogenous leukemia, can be amplified against a background of normal cells using the PCR [30]. Primers anneal to the mRNA produced by the hybrid 9;22 translocation, and amplify only the abnormal RNA.

Inherited Disease

Although most applications of PCR to inherited diseases do not involve archival material, PCR is used extensively to analyze peripheral blood DNA and cells obtained by amniocentesis. Study of tissues may be equally useful. We have used PCR to amplify DNA from the β-globin gene in placental sections containing maternal sickled cells. The fetal genotype for the hemoglobin S gene can be determined by direct gel analysis of a deparaffinized 5-μm section to obtain an amplified product across the sixth codon of the β-globin gene. The amplified product can be cleaved with a restriction enzyme that recognizes the DNA sequence in that region for hemoglobin A, but does

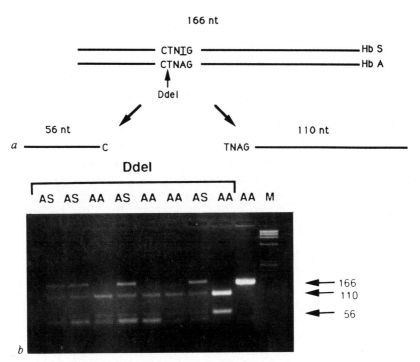

Fig. 9. *a* Amplification across the sixth codon of the β-globin region gives a 166-nt fragment with an internal restriction enzyme site in the gene coding hemoglobin A. The restriction enzyme DdeI will cleave the 166-nt amplified fragment into 56- and 110-nt fragments. The point mutation producing hemoglobin S results in a sequence not recognized by the enzyme; therefore, when the gene coding hemoglobin S is present, there will be no cleavage by DdeI. *b* The direct gel analysis of 5-μm sections of umbilical cords shows a 166-nt band of amplification with genotype AA, which cleaves to 110 and 56 nt with DdeI. However, genotype of AS gives bands of 166, 110, and 56 nt after cleavage with DdeI due to lack of cleavage of one allele by the restriction enzyme.

not recognize the sequence producing hemoglobin S, due to the point mutation (fig. 9). A fetus with an AA genotype has a two band pattern and a fetus with an AS genotype has a three band pattern.

PCR has been used to identify a mislabeled specimen. An obviously mislabeled lymph node with metastatic breast carcinoma was received on the same day as a mastectomy specimen. A polymorphism analysis by PCR on the lymph node and the mastectomy specimen identified the lymph node as belonging to the mastectomy specimen [31].

In situ Hybridization Using PCR

PCR has been used on intact cells, and protocols are being developed to make this technology available. The cells are digested with proteinase K and the PCR is done in a microcentrifuge tube [32]. In situ hybridization is carried out in wells rather than slides, using enzyme-labeled DNA probes since the cells dislodge during the process when doing the PCR on cells fixed onto slides.

Cost of the PCR

The main portion of the cost of the PCR is technical time, because the procedures are much more labor intensive than most current laboratory tests. The approximate cost (1991) for extracting DNA and running a PCR with direct gel analysis of products is $78.05, assuming one specimen per run. This cost will decrease when doing multiple specimens. The cost analysis includes technical time and reagents (Appendix D). Hybridization adds an additional $91.57 per run onto the procedure.

Conclusion

PCR amplification of DNA and RNA in paraffin-embedded tissues has resulted in the availability of a wide variety of material for DNA analysis. Although fixation adversely affects nucleic acids, in most cases amplification can still be achieved from these tissues. The target sequences which can be amplified will be limited only by the imagination of those doing the PCR.

Appendix A – Dealing with Contamination

1   Physically separate the following areas: (1) specimen processing prior to PCR, including DNA extraction; (2) PCR set-up; (3) post-PCR work such as running a gel and blotting.
2   Dismantle and soak pipetters in 20% sodium hypochlorite every other week, and rinse with water (e.g. Millipore Alpha Q Water Purification System). Then air dry. Calibrate pipetters after reassembly.
3   To decrease the risk of contamination from pipetters (which are a major source), use pipet tips with filters (Aerosol Resistant Tips, San Diego, Calif.). Some companies recommend positive displacement pipetters, discarding the barrel after each use.

These are frequently back-ordered and are also very expensive. Pipet tips with filters are a good compromise.

4   Aliquot all reagents for the PCR into small volumes so that tubes are entered a limited number of times, and, if a reagent becomes contaminated, the tube can be discarded without having to discard all of the stock reagent. With each PCR, keep a detailed account of which tube is used so that if contamination occurs, it can be traced more easily.

5   The DNA Carryover Prevention Kit (Cetus Corp., Emeryville, Calif.) can be added to each PCR to guard against cross-contamination from a prior PCR. Using the kit will add $1.50 per tube to the cost of the PCR test.

6   Always use gloves, and a meticulous technique. Even with all of the above precautions, contamination can still occur. It is imperative to run a negative control with each run, and, if the negative control shows a positive result, discard the run and look for the source of contamination.

Appendix B – Rapid Method of Specimen Processing

1   Cutting tissues for the PCR requires meticulous specimen handling from the histology laboratory with knife and instruments. Wipe the instruments and the knife blade with 1% sodium dodecyl sulfate (SDS) followed by 90% ethanol. Between each block cut, wipe the blade again with 1% SDS followed by 90% ethanol. Handle the tissue as little as possible. Include one negative case with each run as a control for possible contamination during specimen processing. This can be used along with the negative controls described in Appendix A, step 6.

2   Place two 10-μm sections of the paraffin-embedded tissue into a 500-μl microcentrifuge tube.

3   Add 400 μl of wax solvent (e.g. xylene) to the sections in the tube.

4   Spin specimen in a vortex and let stand at room temperature for 5 min.

5   Spin 10 min at high speed (we use 16,000 g) and decant by inverting the tube.

6   Repeat steps 3–5.

7   Add 400 μl of absolute ethanol. Spin 10 min and decant.

8   Add 400 μl of 95% ethanol. Spin 10 min and decant.

9   Add 400 μl of 70% ethanol. Spin 10 min and decant.

10 Let dry at room temperature. Place the PCR mixture directly into the microcentrifuge tube containing the tissue and continue for the PCR.

Appendix C – Extraction of DNA from Paraffin-Embedded Tissues

1   Do steps 1–10 above (Appendix B), placing the tissue sections into a 1.5-ml microcentrifuge tube instead of a 0.5-ml tube.

2   Add 400 μl TE (10 mM Tris buffer, pH 8.0, 1 mM EDTA).

3   Add proteinase K to a final concentration of 250 μg/ml, and SDS to a final concentration of 1%. Incubate at 37 °C for 3 h.

4   Add an additional 250 μg/ml of proteinase K (final concentration = 500 μg/ml) and

continue incubation at 37 °C overnight. Longer incubation, with multiple additions of proteinase K, may improve results.

5   Add an equal volume of phenol/chloroform (50/50 mixture), and place the tubes on a rocker for 10 min.
6   Centrifuge for 10 min at high speed (we use 16,000 *g*). Discard bottom layer with a pipette.
7   Repeat steps 5 and 6 twice (a total of 3 phenol/chloroform extractions).
8   Extract by adding an equal volume of chloroform alone. Rock for 10 min and centrifuge as before.
9   Remove the *top* layer (which includes the DNA) and transfer to another 1.5-ml microcentrifuge tube. Discard the bottom layer. Add 2.2 volumes of absolute ethanol and 0.1 volumes of 3 *M* sodium acetate, pH 5.5.
10 Place the specimen in the −70 °C freezer for at least 1 h (overnight or longer may be better).
11 Remove from freezer and centrifuge for 15 min. Decant supernatant. A DNA pellet should be visible at this point.
12 Add 250 µl of 70% ethanol and invert several times. Centrifuge for 15 min.
13 Decant and allow to dry thoroughly while inverted.
14 Resuspend DNA in 100 µl TE and allow to go into suspension at least overnight. Quantitate DNA. We usually use 1 µg of DNA in each PCR reaction, although 500 ng should be sufficient.

Appendix D – Cost of the PCR (One Test per Run)

Procedure	Hours	Technical time, $ ($25/h)	Reagents, $	Total cost, $
Log-in and cut	0.25	6.25	0.00	6.25
Deparaffinization	0.42	10.41	0.00	10.41
Extraction	0.47	11.75	2.30	14.05
Quantitation	0.25	6.25	0.00	6.25
PCR	0.42	10.41	3.84	14.25
Gel electrophoresis	0.37	9.25	2.20	11.45
Southern blot	0.50	12.50	4.70	17.20
Hybridization (^{32}P)	1.25	31.25	43.12	74.37
Result reporting	0.08	2.08	0.00	2.08
Restriction enzyme	0.43	10.83	2.48	13.31

Acknowledgments

I wish to acknowledge the technical assistance and expertise of Solida Mak and Linda Covill, technologists in the Molecular Pathology Laboratory, who are responsible for significant contributions in designing the protocols currently in use in our laboratory. Dr. Greg Buffone, and technologists in his laboratory, should be recognized for contributions to the protocols for PCR and extraction. Dr. Don Singer has provided continued support and invaluable comments regarding this chapter.

References

1   Mullis K: The unusual origin of the polymerase chain reaction. Sci Am 1990;262:56–65.
2   Mullis K, Faloona F: Specific synthesis of DNA in vitro via a polymerase-catalyzed chain reaction. Methods Enzymol 1987;155:335–350.
3   Saiki R, Scharf S, Faloona F, et al: Enzymatic amplification of beta-globin sequences and restriction site analysis for diagnosis for diagnosis of sickle cell anemia. Science 1985;230:1350–1354.
4   Saiki R, Gelfand D, Stoffel S, et al: Analysis of DNA with a thermostable DNA polymerase. Science 1988;239:487–491.
5   Sissons J, Borysiewicz LK: Human cytomegalovirus infection. Thorax 1989;246:241–246.
6   de Franchis R, Cross N, Foulkes N, Cox T: A potent inhibitor of Taq polymerase copurifies with human genomic DNA. Nucleic Acids Res 1988;16:10355.
7   Rogers BB, Josephson SL, Mak SK: Detection of herpes simplex virus using the PCR followed by endonuclease cleavage. Am J Pathol 1991;139:1–6.
8   Sarker G: Shedding light on PCR contamination. Nature 1990;343:27.
9   Wright P, Wynford-Thomas D: The polymerase chain reaction: miracle or mirage? A critical review of its uses and limitations in diagnosis and research. J Pathol 1990;162:99–117.
10 Greer C, Peterson S, Kiviat N, Manos M: PCR amplification from paraffin-embedded tissues. Am J Clin Pathol 1991;95:117–124.
11 Rogers BB, Alpert L, Hine E, Buffone G: Analysis of DNA in fresh and fixed tissue by the polymerase chain reaction. Am J Pathol 1990;136:541–548.
12 Jackson D, Lewis F, Taylor G, et al: Tissue extraction of DNA and RNA and analysis by the polymerase chain reaction. J Clin Pathol 1990;43:499–504.
13 Ben-Ezra J, Johnson D, Rossi J, et al: Effect of fixation on the amplification of nucleic acids from paraffin-embedded material by the polymerase chain reaction. J Histochem Cytochem 1991;39:351–354.
14 Godec M, Asher D, Swoveland P, et al: Detection of measles virus genomic sequences in SSPE brain tissue by the polymerase chain reaction. J Med Virol 1990;30:237–244.
15 Shibata D, Arnheim N, Martin W: Detection of human papilloma virus in paraffin-embedded tissue using the polymerase chain reaction. J Exp Med 1988;167:225–230.
16 Higuchi R: Rapid, efficient DNA extraction for PCR from cells or blood. Amplifications (published by Perkin-Elmer Cetus) 1989(May);2:2–3.
17 Lampertico P, Malter J, Colombo M, Gerber M: Detection of hepatitis B virus DNA in formalin-fixed, paraffin-embedded liver tissue by the polymerase chain reaction. Am J Pathol 1990;137:253–258.
18 Salimans M, van de Rijke F, Raap A, van Elsacker-Niele A: Detection of parvovirus B19 DNA in fetal tissues by in situ hybridisation and polymerase chain reaction. J Clin Pathol 1989;42:525–530.

19 Salimans M, Holsappel S, van de Rijke F, et al: Rapid detection of human parvovirus B19 DNA by dot-hybridization and the polymerase chain reaction. J Virol Methods 1989;23:19–28.

20 Shibata D, Brynes R, Nathwani B, et al: Human immunodeficiency viral DNA is readily found in lymph node biopsies from seropositive individuals. Am J Pathol 1989;135:697–702.

21 Chehab F, Xiao X, Kan Y, Yen T: Detection of cytomegalovirus infection in paraffin-embedded tissue specimens with the polymerase chain reaction. Mod Pathol 1989;2:75–78.

22 Telenti A, Aksamit A, Proper J, Smith T: Detection of JC virus DNA by polymerase chain reaction in patients with progressive multifocal leukoencephalopathy. J Infect Dis 1990;162:858–861.

23 Rouah E, Rogers BB, Wilson D, et al: Demonstration of Epstein-Barr virus in primary central nervous system lymphomas by the polymerase chain reaction and in situ hybridization. Hum Pathol 1990;21:545–550.

24 Nuovo G: Human papillomavirus DNA in genital tract lesions histologically negative for condylomata. Am J Surg Pathol 1990;14:643–651.

25 Wan J, Trainor K, Brisco M, Morley A: Monoclonality in B cell lymphoma detected in paraffin wax embedded sections using the polymerase chain reaction. J Clin Pathol 1990;43:888–890.

26 Frye R, Benz C, Liu E: Detection of amplified oncogenes by differential polymerase chain reaction. Oncogene 1989;4:1153–1157.

27 Bianchi A, Navone N, Conti C: Detection of loss of heterozygosity in formalin-fixed paraffin-embedded tumor specimens by the polymerase chain reaction. Am J Pathol 1991;138:279–284.

28 Yandell D, Campbell T, Dayton S, et al: Oncogenic point mutations in the human retinoblastoma gene: their application to genetic counseling. N Engl J Med 1989;321:1689–1695.

29 Rodenhuis S, van de Wetering M, Mooi W, et al: Mutational activation of the K-ras oncogene. N Engl J Med 1987;317:929–935.

30 Morgan G, Janssen J, Guo A-P, et al: Polymerase chain reaction for detection of residual leukaemia. Lancet 1989;i:928–929.

31 Shibata D, Namiki T, Higuchi R: Identification of a mislabeled fixed specimen by DNA analysis. Am J Surg Pathol 1990;14:1076–1078.

32 Amplifications. 1990(March);4:20–21.

Beverly B. Rogers, MD, Department of Pathology, Women and Infant's Hospital, 101 Dudley Street, Providence, RI 02905 (USA)

Garvin AJ, O'Leary TJ, Bernstein J, Rosenberg HS (eds): Pediatric Molecular Pathology: Quantitation and Applications. Perspect Pediatr Pathol. Basel, Karger, 1992, vol 16, pp 120–133

A Natural Immune System in Pregnancy Serum Lethal to Human Neuroblastoma Cells: A Possible Mechanism of Spontaneous Regression

Robert P. Bolande

Department of Pathology, East Carolina University School of Medicine, Greenville, N.C., USA

Introduction

Neuroblastoma (NB) is a unique neoplasm. When it appears after 1 year of age, it is very aggressive and quite unresponsive to therapy. Before 1 year of age, it is likely to undergo 'spontaneous' regression either by cytolysis or cytodifferentiation into ganglioneuroma. Before 1 year of age regression occurs in 45% of cases. In the congenital disseminated form commonly referred to as stage IV-S, regression may occur in 90% of the cases [1–4].

The 'spontaneous' regression of cancer refers to the healing or disappearance of malignant growth in the absence of significant medical intervention. Estimates of its occurrence rate are now placed at between 1:80,000 and 1:100,000 cancers at all ages [5], but the reported cases do not approximate the true incidence of this phenomenon. The natural history of untreated cases of cancer is rarely reported or discussed, the assumption being that the course of untreated cancer is always relentless and fatal. The regression of incipient or preclinical cancer may indeed be common, but evidence of its occurrence may be subtle, or no trace may be left. Spontaneous regression is most often reported in cases of renal cell carcinoma, malignant melanoma, and uterine choriocarcinoma, which are usually metastatic at the time of diagnosis [6]. Under these conditions, oncotherapy may be considered futile and withheld. To a lesser extent, spontaneous regression has been observed in embryonal carcinoma of the testis, bladder carcinoma, bronchogenic carcinoma, and breast cancer [6].

The outstanding, spontaneously regressing cancer is infantile NB [4]. Malignancy in newborn and young infants is rare. Bader and Miller [7] showed that malignant neoplasms occur at annual rates of $183.4/10^6$ in infants under 1 year of age, and $36.5/10^6$ newborns under 1 month. Half are found at birth and two thirds during the first weeks of life. The most common tumors found at this time are NB, leukemia, and renal tumors. NB is prevalent, occurring at annual rates of $62.7/10^6$ infants under 1 year of age and $19.7/10^6$ infants under 1 month of age. Kidney tumors (Wilms' tumor and congenital mesoblastic nephroma) occur in $21.6/10^6$ infants under 1 month and $3.6/10^6$ at 1 year of age. It is remarkable that in infants under 1 year of age, the incidence of cancer is 3 times greater than its mortality. This ratio of incidence to mortality is 8:1 for NB in the 1st year, 5.4:1 for kidney tumors, but only 1.5:1 for leukemia. In retinoblastoma, the incidence is 159 times greater than the mortality. The disparity between incidence and mortality may, in part, reflect the efficacy of improved therapy, particularly in the kidney tumors and retinoblastoma. It could also reflect the essential benignity of the neoplasms or their tendency, in the early months of life, for regression and cytodifferentiation into benign forms during the so-called 'oncogenic grace period' [8]. This is clearly the case with kidney tumors in infancy, in which the great majority are the benign congenital mesoblastic nephroma rather than true Wilms' tumor [9, 10]. As there have been no significant advances in the therapy of NB, the enhanced survivorship of infants with this disease would seem to reflect its regressive tendencies.

The phenomenon of spontaneous regression of NBs has intrigued oncologists and pathologists for many years. Experimental studies as to its possible mechanisms are limited. Outstanding work in this field was carried out by Hellström and co-workers [11, 12]. They showed that the lymphocytes of patients with NB were cytotoxic to NB cells, but that blocking serum antibodies to tumor antigens often masked these antigens and thus protected the cells from the cytotoxic effect of these lymphocytes. They concluded that spontaneous regression was probably due to immune mechanisms.

Miller [13] observed that NB scarcely exists in Uganda, Nigeria, much of Africa, and parts of India and Puerto Rico. He correlated this low incidence with maternal hyperimmunoglobulinemia due to the prevalence of malaria and other parasitic infections, which may exert a transplacental 'anticarcinogenesis' effect that inhibits the development of NB. In the light of these observations and experiences, we set out to define immune mechanisms that might be responsible for the regression of NB.

Experimental Studies

As known since the early part of the century, fresh human serum, particularly from pregnant women, is cytolytic to xenogeneic and certain tumor cells [14–17]. Relatively little recent work has been done in this field, as the concept of humoral immunosurveillance against cancer was downplayed and overshadowed by studies emphasizing cellular immunity. Since a natural IgM antibody, in conjunction with complement, is responsible for the cytolytic reaction, mainly in murine cancer cells [18–23], we speculated that this system might be operative in the regression of certain human NBs. The following is a synopsis of the experiments testing this hypothesis [24–26].

Murine Cancer Cells

It was first necessary to further investigate and analyze the cytotoxic system for murine cells [24]. Using L cells, Sarcoma 180, and the Ehrlich ascites tumor, the cytotoxic activity was universally present in the third trimester pregnancy sera but was not limited to pregnancy sera, as about 30% of male and nonpregnant female sera were also cytotoxic.

Serum cytotoxicity was manifested as cell death, which occurred within 15 min of incubation at 37 °C and proceeded to complete cytolysis. It was initiated within 5 min by the ultrastructural appearance of membrane 'pores' and discontinuities (fig. 1), rapidly progressing to complete membrane loss and a lethal increase in cell permeability. Increased membrane permeability was the basis for the uptake of trypan blue, used as a quantitative assay of cell death.

Cell death was mediated by complement and involved both the classical and alternative pathways (fig. 2). Loss of cytotoxicity occurred when serum was heated to 56 °C or treated with ethylenediaminetetraacetic acid (EDTA). The latter chelates Ca^{2+} and Mg^{2+}, upon which both the classical and alternative pathways of complement depend.

The activation of complement followed the binding of IgM to the target cells' surfaces, which in turn sensitized and amplified their reactivity to serum complement [25] (fig. 3). This was shown by a series of immunologic and immunochemical experiments: (1) Adsorption of serum with tumor cells at 4 °C produced sensitized cells whose reactivity to dilute serum complement was amplified 16–32 times. The sensitization of cells took place rapidly, beginning within 5 min of adsorption and completed within 30 min. (2) Immunoperoxidase staining of sensitized cells showed diffuse surface

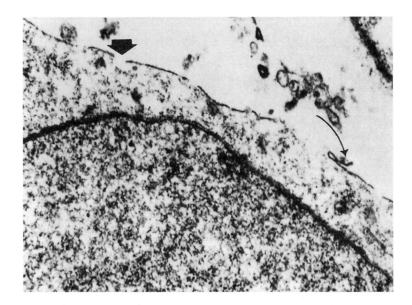

Fig. 1. Electron microscopy of early cytotoxicity on L cell after 5 min in pregnancy serum at 37 °C. Discontinuities or pores (solid arrow) are present along the cell membrane with focal rolling back of the cell membrane (curved arrow). Cytoplasmic debris of lysed cells is seen in background. ×20,475.

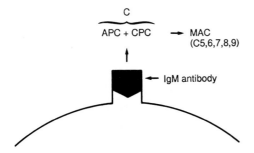

Fig. 2. The cytolytic reaction. Both the alternative and classical pathway of complement are activated by attachment of natural IgM antibody to the cell surface. The membrane attack complex causes the membranolysis (fig. 1) followed by cytolysis.

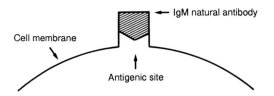

Fig. 3. The sensitized cell. Adsorption of cells with pregnancy serum or IgM at 4 °C produces a 16–32× amplification of the cells' reactivity to serum. This is due to the saturation of cell surface antigenic sites with IgM.

binding of IgM, whereas little or no demonstrable IgG or IgA. (3) The lytic action of complement could be blocked by treating sensitized cells with monoclonal anti-human IgM, whereas anti-IgA and anti-IgG had no effect. (4) The sensitizing activity was localized to a chromatographically purified IgM fraction of pregnancy serum (fig. 4).

Human Neuroblastomas

Twenty-six human cancer cell lines were surveyed for evidence of cytotoxicity in response to normal pregnancy serum [26]. Cytolytic activity occurred in 4 out of 8 NB cell lines studied, but none of the other human cancers (table 1). We analyzed the nature of this cytotoxicity using the murine experiments as a template (table 2). NGP and SK-N-SH cells were the most reactive, followed by LAN-5 and IMR-32. These cell lines were all established from widely metastatic (stage IV) NBs. A cell line derived from a metastasis of a stage IV NB in a 3-year-old child showed marked neuron-specific enolase activity, catecholamine production, and amplification of N-*myc* (fig. 5). Upon re-establishment of the tumor in culture after passage through the mouse, the cytotoxic and immunological reactivity remained the same as in the original culture.

Preincubation of reactive NB cells SK-N-SH and NGP with pregnancy serum or purified IgM at 4 °C showed about the same degree of sensitization to serum complement. Male serum and cord serum showed a markedly reduced cytolytic activity on cells presensitized with pregnancy serum or IgM. The lytic action of complement was blocked by incubating these sensitized NB cell lines with monoclonal anti-human IgM antibody but was not blocked by incubation with anti-IgG or anti-IgA (table 3).

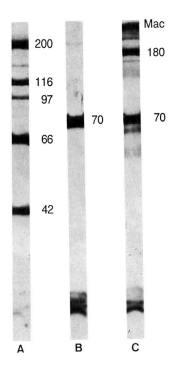

200

116
97

66

42

70

Mac
180

70

A   B   C

Fig. 4. Electrophoretic pattern of chromatographically prepared pregnancy IgM by Sephadex G-200 column. Bio-Rad silver stain. Lane A shows the marker proteins. Molecular weights (in kD) are indicated to the right. Lane B is commercially purified IgM. Only the 70-kD chain of IgM is seen. Lane C is the pattern derived from the chromatographically prepared pregnancy IgM. The 180-kD line is the monomeric form of IgM. The 70-kD line is the chain. The line labeled Mac represents macroglobulins. After the macroglobulins were eliminated by immune precipitation with antimacroglobulin, the preparation showed a 20-fold increase in the IgM/protein ratio over normal serum.

By immunoperoxidase stains, sensitized SK-N-SH and NGP cells showed marked diffuse deposition of IgM on their surfaces (fig. 6). Little or no IgG or IgA was detectable. Weakly reactive or nonreactive NB cells showed no deposition of IgM, IgA, or IgG. The powerful cytolytic system operative on murine cells was obviously the same for NB.

Characterization of the NB Antigen [27]

NB cell wall preparations were solubilized with an amphoteric detergent (3-[(3-cholamidopropyl)dimethylammonio]-1-propanesulfonate) (CHAPS).

Table 1. Human NBs and other human cancer cell lines

Tumor	Source	Code names	Cytotoxicity of undiluted serum % dead cells
NB	ATCC	SK-N-SH	40
	G.B.[a]	NGP	40
	NCI[b]	LAN-5	30
	ATCC	IMR-32	30
	G.B.[a]	NLF	0
	ATCC	SK-N-ME	0
	NCI	SK-N-LE	0
	NCI	SK-N-DZ	0
Rhabdomyosarcoma	ATCC	RD-136	0
	ATCC	A-204	0
Wilms' tumor	ATCC	SK-NEP-1	0
	ATCC	Wiltu-1	0
	Primary	Wiltu	0
Choriocarcinoma	ATCC[c]	JEG-3	0
Renal cell carcinoma	ATCC	A-704	0
	ATCC	A-498	0
Melanoma	ATCC	SK-MEL-28	0
	ATCC	SK-MEL-3	0
	ATCC	HT-144	0
Osteogenic sarcoma	ATCC	MNNG/HOS/CLFS	0
Ewing's sarcoma	ATCC	SK-ES-1	0
Lung sarcoma	ATCC	CAL-U-6	0
Breast carcinoma	ATCC	SK-BA-3	0
Liver carcinoma	ATCC	SK-HEP-1	0
Glioblastoma III	ATCC	U-87-76	0
	ATCC	U373-46	0

[a] Dr. Garrett Brodeur, Washington University, St. Louis, Mo.
[b] Dr. Mark Israel, National Cancer Institute, Bethesda, Md.
[c] American Tissue Culture Collection.

Table 2. Sensitization of NBs SK-N-SH and NGP with various sera and pregnancy IgM

Cell lines	Sensitized with	Cytotoxicity of pregnancy serum % dead cells	Cytotoxicity of male serum % dead cells
SK-N-SH	Pregnancy serum	85	5
	Pregnancy IgM	70	–
	Male serum	30	5
	Cord serum	30	5
	PBS control	5	0
NGP	Pregnancy serum	80	10
	Pregnancy IgM	75	–
	Male serum	25	5
	Cord serum	15	0
	PBS control	5	5

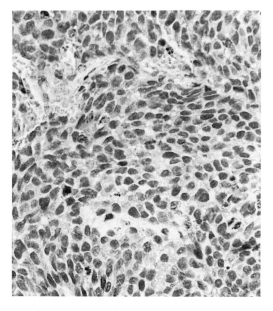

Fig. 5. Subcutaneous implant of NGP cells in nude mouse (BALB/c ANCrl:nu/nu). This tumor, which developed 8 weeks after the subcutaneous injection of 4×10^6 NGP cells from a cell line derived from a metastasis of a stage IV NB in a 3-year-old child, shows a highly anaplastic small cell tumor with numerous mitoses and a suggestion of lobulation. HE. $\times 320$.

Table 3. Blocking experiments with monoclonal anti-Ig antibodies on sensitized NB cells

NB cell lines	Cytotoxicity of undiluted serum % dead cells
Sensitized SK-N-SH	75
+ anti-IgM (1/20)	20
+ anti-IgG (1/20)	75
+ anti-IgA (1/20)	70
+ anti-Macl (1/20)	65
+ PBS	0
Unsensitized SK-N-SH	35
Sensitized NGP	90
+ anti-IgM (1/20)	40
+ anti-IgG (1/20)	90
+ anti-IgA (1/20)	90
+ anti-Macl (1/20)	85
+ PBS	0
Unsensitized NGP	40

Concentrations of these preparations demonstrated that NGP and SK-N-SH extracts were immunoreactive with pregnancy IgM using the dot-blot technique (fig. 7). Nonreactive, control NB cell lines were not immunoreactive. The dot-blot immunoreactivity was trypsin-labile but unaffected by periodate. Using the Western blot technique, the immunoreactivity of this antigen in NGP cells appears to be related to a protein with a molecular weight of about 50 kD. Lectins, such as concanavalin A and phytohemagglutinin, as well as glucosidase and galactosidase would have blocked the cytolytic action if the antigenic component had been a polysaccharide. Failure of these lectins to block the cytolysis excludes the possibility that the antigen is a polysaccharide such as an HLA or a blood group substance.

Discussion

A cytotoxic system in pregnancy serum is capable of destroying NB cells to the exclusion of other human cancers tested. It consists of a 'natural' IgM antibody that binds to the cells' surface and amplifies their lytic response to serum complement. It is similar to the system involved in the lysis of murine cancer cell lines.

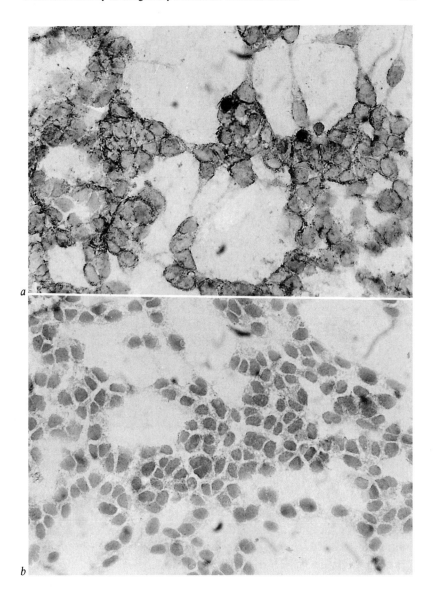

Fig. 6. Immunoperoxidase reaction for IgM on NB cells. *a* Surface deposition of IgM on sensitized NGP cells. Immunoperoxidase stain-goat anti-human IgM conjugate. ×320. *b* Absence of surface deposition of IgA on sensitized NGP cells. Immunoperoxidase stain-goat anti-human IgA conjugate. ×320.

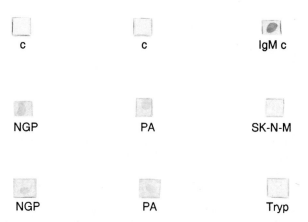

Fig. 7. Dot-blot preparations of cell membrane extracts of NB cells treated with pregnancy IgM and stained with anti-IgM immunoperoxidase conjugates. c = Cell extracts not treated with IgM are negative; IgMc = purified IgM used as a positive control. Extracts of NGP are immunoreactive after IgM treatment. SK-N-MC, a nonreactive NB cell line extract, shows no IgM uptake. In the lower row, NGP extract first treated with trypsin (Tryp) shows loss of binding of IgM; this indicates the protein nature of the antigen. Treatment with periodic acid (PA) has no effect indicating that the antigenic determinant of the reaction is not polysaccharide.

NB cell lines appear to be unique in their reactivity to pregnancy serum when compared with other human cancers. Humoral immune systems have been described for other human cancers of neurocristopathic origin, such as melanoma and certain gliomas [28–31]. This reactivity is due to so-called neuroectodermal-differentiation antigens. It would seem unlikely that such antigens are involved in the cytolytic system described here, as the melanomas and gliomas tested did not react to gestational serum.

Cell sensitization would seem to result from the saturation of antigenic binding sites with IgM. While adequate levels of IgM and complement are required for the lytic action to proceed, the sensitizing capability of serum does not correlate with serum IgM levels.

Studies of the serum of pregnant and nonpregnant women and of men show that the sensitizing activity of pregnancy serum is greatest and universally present in the third trimester; it occurs with low or normal IgM levels. High levels of IgM did not enhance the cytolytic activity. In the nonpregnant groups, there was less activity even with very high IgM levels [32].

A direct transplacental transfer of a gestational IgM antibody is unlikely, as IgM does not cross the placenta. A natural IgM antibody, occurring in gestational serum, might well protect the mother from transplacental NB metastases. But how could this immune system protect the fetus and newborn infant? Since maternal lymphocytes are normally transferred transplacentally to the fetus [33–35], enough maternal B cells might gain access to the fetal and newborn circulation where they could transiently produce sensitizing levels of the natural antibody. Present studies of infants in the first year of life have shown high levels of sensitizing antibodies in the presence of very low serum IgM levels [32]. It is therefore possible, through IgM antibody production by a small population of these B lymphocytes, that the pregnant woman confers anti-NB humoral immunity on her infant. This effect seems to persist through the first months of life. With a deficiency or absence of such a system in the infant, malignant neuroblasts could proliferate and progress to a neuroblastoma with impunity.

The antigen is expressed in half of the NB cell lines studied. Tumor antigens are known to be shed or lost in vitro. It is thus important to study primary explants or suspensions of fresh NBs to see if the antigen is more universally expressed in vivo to determine whether there may be a place for the treatment of NB with fresh, third trimester pregnancy serum or with concentrates of pregnancy IgM.

References

1   Evans AE, Gerson J, Schnaufer L: Spontaneous regression of neuroblastoma. Natl Cancer Inst Monogr 1976;44:49–54.
2   Pochedly C: Neuroblastoma in infancy; in Pochedly C (ed): Neuroblastoma. Acton, Mass., Publishing Sciences Group, 1976, pp 22–23.
3   Bolande RP: Developmental pathology. Am J Pathol 1979;94:627–684.
4   Cole WH: Spontaneous regression of cancer and the importance of finding its cause. Natl Cancer Inst Monogr 1976;44:5–9.
5   Franklin CIV: Spontaneous regression of cancer; in Stoll BA (ed): Prolonged Arrest of Cancer. Chichester, Wiley 1982, pp 103–116.
6   Bashford EF: The spontaneous regression of hypernephroma. Am J Cancer 1935;24: 839.
7   Bader JL, Miller RW: US cancer incidence and mortality in the first year of life. Am J Dis Child 1979;133:157–159.
8   Bolande RP: Spontaneous regression and cytodifferentiation of cancer in early life: The oncogenic grace period. Surv Synth Pathol Res 1985;4:296–311.
9   Bolande RP: Congenital mesoblastic nephroma. Perspect Pediatr Pathol 1973;1: 227–250.
10 Bolande RP: Congenital and infantile neoplasia of the kidney. Lancet 1974;ii:1497–1499.
11 Hellström KE, Bill AH, Pierce GE, et al: Studies on cellular immunity to human neuroblastoma cells. Int J Cancer 1970;6:172–188.

12   Hellström KE, Hellström I: Spontaneous tumor regression: Possible relationships to
     in vitro parameters of tumor immunity. Natl Cancer Inst Monogr 1976;44:5–9.
13   Miller RW: Ethnic differences in cancer occurrence: Genetic and environmental
     differences with particular reference to neuroblastoma; in Mulvihill J, Miller R,
     Fraumeni J (eds): Genetics of Human Cancer. New York, Raven Press, 1977, pp 1–
     14.
14   Wells HG: Chemical Pathology. Philadelphia, Saunders, 1925.
15   Glumac G, Mates A, Eidinger D: The heterocytotoxicity of human serum. III.
     Studies of the serum levels and distribution of activity in human populations. Clin
     Exp Immunol 1976;26:601–608.
16   Bolande RP, Todd EW: The cytotoxic action of normal human serum on certain cells
     propagated in vitro. Arch Pathol 1958;66:720–732.
17   Bolande RP: Cytotoxic action of human serum on atypical mammalian cell lines.
     Lab Invest 1960;9:475–489.
18   Eidinger D, Gery I, Bello E: The heterocytotoxicity of human serum. II. The role of
     natural antibody and of the classical and alternative complement pathways. Cell
     Immunol 1977;29:187–194.
19   Laskov R, Simon E, Gross J: Studies on a naturally occurring antibody against
     mouse Landschutz ascites tumor cells. Cytotoxicity assay and quantitation in normal
     sera. Int J Cancer 1968;3:96–105.
20   Laskov R, Simon E, Ram D, et al: Studies of a naturally occurring human antibody
     against mouse Landschutz ascites tumor cells. II. Identification of the antibody as
     IgM and its partial purification. Int J Cancer 1968;3:107–115.
21   Laskov R, Simon E, Gross J: Studies of a naturally occurring human antibody
     against mouse Landschutz ascites tumor cells. III. The nature of the antigenic site on
     the tumor cell. Int J Cancer 1968;3:511–522.
22   Violand BN, Racker E: Permeabilization of Ehrlich ascites tumor by human serum.
     Cell Immunol 1981;61:213.
23   Gross J, Simon E, Szwarc-Bilotynski L, et al: Complement-dependent lysis of Ehrlich
     ascites cells by human serum (ascitolysin) is lowered in cancer patients and raised in
     pregnant women. Eur J Cancer Clin Oncol 1986;22:13–19.
24   Bolande RP, Arnold JL, Mayer DC: Natural cytotoxicity of human serum. A natural
     IgM 'antibody' sensitizes transformed murine cells to the lytic action of complement.
     Pathol Immunopathol Res 1989;8:46–60.
25   Bolande RP, Mayer DC: Isolation of a natural cytotoxic IgM 'antibody' in human
     serum sensitizing L cells to complement. Proc Soc Exp Biol Med 1989;191:387–390.
26   Bolande RP, Mayer DC: The cytolysis of human neuroblastoma cells by a natural
     IgM 'antibody' complement system in pregnancy serum. Cancer Invest 1990;8:603–
     611.
27   Bolande RP, Mayer DC: Characterization of neuroblastoma antigen reactive with
     natural IgM antibody in gestational serum (in preparation).
28   Houghton AN, Taormina MC, Ikeda H, et al: Serological survey of normal humans
     for 'natural' antibody to cell surface antigens of melanoma. Proc Natl Acad Sci USA
     1980;77:4260–4264.
29   Carrel S, de Tribolet N, Mach J-P: Expression of neuroectodermal antigens common
     to melanomas, and gliomas and neuroblastomas. I. Identification by monoclonal
     anti-melanoma and anti-glioma antibodies. Acta Neuropathol (Berl) 1982;57:158–
     164.

30 Kornblith PL, Coakham HB, Pollock LA, et al: Autologous serologic responses in glioma patients. Cancer 1983;52:2230–2235.
31 Matsutani M, Suzuki T, Hori T, et al: Cellular immunity and complement levels in host with brain tumors. Neurosurg Rev 1984;7:29–35.
32 Bolande RP, Mayer D: Seroepidemiological survey of natural antineuroblastoma antibody and IgM levels in pregnant women and infants (in preparation).
33 Beer AE, Billingham RE: Maternally acquired runt disease. Immune lymphocytes from the maternal blood can traverse the placenta and cause runt disease in the progeny. Science 1973;179:240–243.
34 deGrouchy J, Trebuchet C: Etude des lymphocytes foetaux dans le sang de la femme enceinte. Bull Eur Soc Hum Genet 1970;4:65.
35 Walknowska J, Conte FA, Grumbach MM: Practical and theoretical implications of fetal/maternal lymphocyte transfer. Lancet 1969;i:1119–1122.

Robert P. Bolande, MD, Department of Pathology, East Carolina University School of Medicine, Greenville, NC 27858–4354 (USA)

Garvin AJ, O'Leary TJ, Bernstein J, Rosenberg HS (eds): Pediatric Molecular Pathology:
Quantitation and Applications. Perspect Pediatr Pathol. Basel, Karger, 1992, vol 16, pp 134–159

Quantitative Methods in Pediatric Pathology

An Overview with Special Consideration of
Developmental Renal Abnormalities

Karen Schmidt, Carlo Pesce[1]

Laboratory of Pathology, NCI, and Renal Cell Biology Section,
NIDDK, Bethesda, Md., USA

General Principles of Image Analysis

Quantitation as a Supplement to Visual Recognition

A common drawback of all microscopic methods is their dependence on
visual recognition. The traditional approach relies on the subjective appreciation of geometrical or physical characteristics, for instance number,
shape, dimension, and intensity of staining. Human visual perception is
relatively better at identifying patterns than at evaluating differences in
amount and may be easily mistaken in discerning shape.

Another limitation of visual recognition is its reliance on two-dimensional histologic preparations. Pathologists routinely make assumptions
about the real, i.e. three-dimensional, structure by comparing the two-
dimensional specimen examined with an ideal control [1]. Satisfactory three-
dimensional reconstruction (3DR) of abnormal structures to represent the
control has occasionally been performed, sometimes with unexpected results.

An exclusively visual approach to morphologic evaluation extracts only
part of the information contained in the specimen. The adoption of quantitative methods could considerably improve the yield and accuracy of the
results [2, 3]. The differences between visual and morphometric evaluation
are evident when they are systematically compared. For instance, peripheral
muscle atrophy was recognized in 50% of specimens by morphometry, but

---

[1] We are most grateful to Dr. Liliane Striker and Dr. Gary Striker for their sugges-
tions.

only in 15% of the same specimens with nonquantitative visual evaluation [4]. Quantitative data can be analyzed statistically, allowing objective conclusions to be drawn. These two steps, acquisition of quantitative data and their statistical evaluation, are strictly related and interdependent features. Additional procedures applied to image-analysis data include 3DR and allometry.

As a general observation, the scope of the analysis in pediatric and developmental conditions may require consideration of tissue growth. Methods of image acquisition and processing are particularly helpful in addressing growth problems. The first part of this article is an overview of the main quantitative methods used in morphologic analysis and their recent applications to pediatric pathology. The second considers the quantitative approach to the study of several pediatric renal diseases.

Classic Morphometric Methods

Volume fraction (Vv), surface area per volume (Sv), and length per volume (Lv) are three basic morphometric descriptors that relate geometric characteristics of the structure examined to the reference space. They are traditionally calculated with point or line counting by using simple equations expressing density and distribution of the structure in a given plane of section, which is viewed as an unbiased sample of the reference space. Vv is a dimensionless variable representing the percentage of space occupied by the structure; it equals the surface fraction, i.e., the percentage of random points superimposed on the plane that hit the structure (fig. 1). Sv, expressed as unit of length^{-1}, relates the surface of the structure to the reference space and is derived from the relationship between the number of intersections of the structure with a line grid of known length (fig. 2). Lv, expressed as unit of length^{-2}, relates the length of a linear structure to the reference space, and is derived from the number of cross sections of the linear structure in a given area (fig. 3). Details of the procedures used to determine these values can be found in several excellent textbooks of morphometry [5–8]. Estimates of relative volume, surface, and length are often calculated with computer-based planimeters; automation allows faster data collection and higher precision of measurements.

Applications of Vv, Sv, and Lv and Derived Functions

Vv, Sv, and Lv may be used to compare changes in structures pertaining to the same reference space or combined to study the configuration of the same structure or to derive mathematically other geometric descriptors.

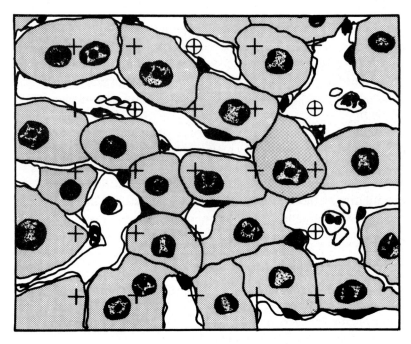

Fig. 1. The classic morphometric approach to the evaluation of volume fraction (Vv) relies on the equation $Vv = Pp$, where Pp is the percentage of points that hit a random section of the structure. In this example, a grid of points is superimposed on a liver section in order to estimate the Vv of the sinusoids. The number of points falling on the sinusoids is recorded. The procedure is repeated several times in order to yield a significant value. It may be noted that the arrangement of the sinusoids is nonrandom, since they converge onto the central lobular vein. The bias due to such anisotropism could be obviated by using a grid with randomly distributed test points. The above issues of significance and anisotropism are discussed in the section 'Usefulness of Vv Measurements'.

Unlike the parent functions, the new descriptors may be absolute, i.e. independent of the reference space, as in total surface area or thickness of biologic laminas.

Examples: Volume fraction measurements are an essential part of the examination of bone biopsies for nonneoplastic disorders [9]. The parameters describing mineralization include extension of the mineralization front (normal value (n.v.), more than 80% of the osteoid surface) and mineralization rate (n.v., 1 µm/day by tetracycline double-labeling). Bone turnover is evaluated through the age-related osteoid index (osteoid volume/osteoid surface, expressed as percentage), density of birefringent lamellae (n.v., < 4), and osteoid fraction of the trabecular surface (n.v., 1%). Such methods have allowed

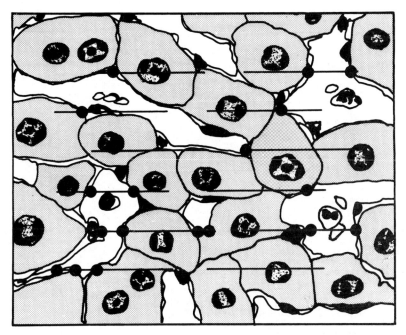

Fig. 2. A series of lines of known length is superimposed on the same picture shown in figure 1 to estimate the Sv of the sinusoids. The number of transections of the test lines with the sinusoidal border is recorded. The procedure is repeated enough times to yield a significant value. The issue of anisotropism can be obviated with different (e.g., curved) line outlays. In contrast to the method illustrated in figure 1 for Vv, an absolute measure, the combined length of the test lines, is introduced in the calculation. As a result, the Sv is a dimensional (unit of length^{-1}) function.

the definition of the amount of renal osteodystrophy in hemodialyzed patients, i.e., decreased trabecular bone fraction and increased osteoid and bone resorption.

In a series of testicular biopsies to determine the effect of varicocele, damage to the seminiferous line was inconsistent, whereas the Vvs of the vessels and of the interstitium and the Leydig cell count were significantly increased in comparison with those of controls. These data suggest that histologic changes in the interstitium could represent the main early diagnostic features of varicocele [10].

The total volume of the lamina propria does not differ significantly in children affected by celiac disease, whether untreated, treated, or undergoing gluten challenge, in comparison with normal controls. Only in patients with untreated celiac disease or with gluten challenge, however, the volume of the surface epithelium decreases and that of the crypts increases, with an excessive cell turnover accounting for this disproportion [11].

In patients with Wilson's disease, liver biopsies showing the characteristic features of the disease had an increase in the Vv and the Sv of the mitochondria, and in the Sv of the

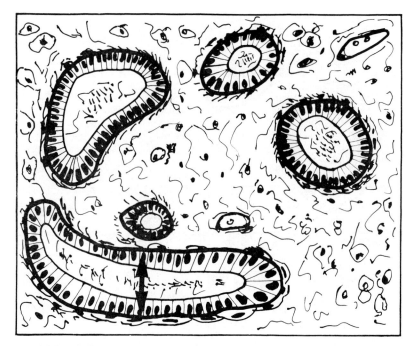

Fig. 3. The Lv of a linear structure (in this case, the epididymis) is estimated from the number of intersections of the structure in the testing area. The epididymis can be considered linear for practical purposes, since its length is not comparable to its diameter. Lv can be used in conjunction with other functions to derive the absolute length. For the epididymis, which is a single structure, the absolute length corresponds to its real length; for multiple linear structures (e.g., seminiferous tubules, renal tubules), the absolute length corresponds to their combined lengths. The absolute measure introduced for computation is the surface of the test area (unit of length^2); Lv is expressed as unit of length^{-2}. The segment with terminal arrowheads representing the perpendicular to a longitudinal section of the tubule constitutes a good approximation of the epididymal diameter.

mitochondrial external membranes as compared with normal controls, although with considerable individual variation [12].

Since the Vv, Sv, and Lv of the sinusoids are increased in cirrhosis, sinusoidal changes should not contribute to the development of portal hypertension [13]. The variations in the Vv, Sv, and Lv of the lumina in subcutaneous hemangiomas suggest a continuum rather than a clear-cut separation between capillary and cavernous types [14]. The length per unit area (La) of bulbar conjunctival vessels was measured in diabetic and nondiabetic subjects from 11 to 68 years of age, divided into 15-year cohorts. The La of vessels measuring less than 30 µm, i.e. capillaries and postcapillar-

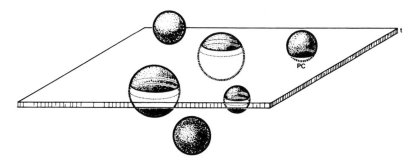

Fig. 4. Calculating the numerical density of particles (Nv) from the profiles on a given area of a histological section requires knowledge of the diameter and shape of the particles. A more refined estimate would also correct for polar caps (PC) and section thickness (t), if t is comparable with the size of the particles, as in the case of these ideal spheres. Slightly ellipsoidal particles can be considered spherical in practice. If the particle shape is considerably different from a sphere, a shape coefficient should be included in the calculation.

ies, decreased with age more in the diabetics than in the nondiabetics; La did not decrease in larger vessels [15].

The increased total area of the carotid body in sudden infant death syndrome, cystic fibrosis, and cyanotic heart disease suggested a relationship with chronic hypoxia [16].

The volumes of the thyroid follicles and their epithelial surfaces, derived from the follicular Vv and Sv measured by point and line counting, increased consensually until age 20 years [17].

Another important quantitative parameter, besides Vv, Sv, and Lv, is the number of objects per unit volume or object density (Nv, expressed as unit of length^{-3}). Nv can be estimated morphometrically from the number of profiles per unit area of the histologic section (fig. 4). This estimate requires knowledge of the thickness of the histologic section, of the dimension of the objects, and often of their shape. If the diameter of the objects is large enough not to interfere with section thickness and the objects are approximately spherical, Nv can be derived from the number of profiles per unit area (Na) divided by the mean diameter of the objects (not of the section profiles). If section thickness is expected to interfere considerably, the computing formula should include some correction for section thickness and 'optically lost caps', i.e. underestimation due to unrecognized polar sections. For nonspherical objects of known shape, Nv can be derived from Na and Vv with the introduction of a shape coefficient.

Usefulness of Nv Measurements

The mean perimeter and mean area of the biliary canaliculi and the Na of their microvilli were assessed in electron micrographs of liver biopsies in arteriohepatic dysplasia and compared with measurements in idiopathic intrahepatic cholestasis, extrahepatic biliary atresia, and noncholestatic liver disease [18]. No differences in the number and distribution of canalicular profiles were found in the four groups or between periportal and centrilobular areas. Analysis of variance showed that in arteriohepatic dysplasia the dimensions of the biliary canaliculi were similar to those in extrahepatic cholestasis and noncholestatic disease, whereas intrahepatic cholestasis was associated with significant canalicular dilatation. No differences in the Nas of microvilli were detected in the four groups, a finding which contrasted with the impression of direct, nonmorphometric observations.

Seshadri et al. [19] showed by discriminant analysis that mean nuclear area and the percentage of cells with nucleoli were the best variables to use in the subtyping of acute childhood lymphoblastic leukemia by the French-British-American (FAB) classification. The area and perimeter of the nuclei and the area and Na of nucleoli were assessed with a computer-based image analysis system. Similar results were obtained by Tosi et al. [20] in a retrospective study of acute lymphoblastic leukemia survival. The cell area was greater and the nuclear/cell area ratio lower in short-term survivors. These data confirm the prognostic significance of the morphologic parameters in the FAB classification and the degree of accuracy that can be achieved with quantitation.

Ouvrier et al. [21] have measured by computer-based planimetry the endoneurial area of sural nerve fascicles, the Na of myelinated fibers, and their external diameter. Myelinated fibers decreased from $18.34 \cdot 10^3/mm^2$ at 1 day to $7.11 \cdot 10^3/mm^2$ at 10 years, showing a parallel increase in size and transverse area of the nerve fascicles with age. In sural nerve biopsies from patients with Friedreich ataxia, aged 3–25 years, the mean density of fibers was lower than normal and decreased progressively with age [22]. Although the total density was normal, large diameter fibers were already scanty in the younger patients.

Size of Cylinders

The diameters of cylinders, such as seminiferous tubules or skeletal muscle fibers, are best measured in the perpendicular to the longest profile in longitudinal sections or to the largest profile in transverse sections ('largest-smallest diameter') (fig. 3). The perpendicular is the better estimator of diameter because it avoids the chords in bent cylinders that are longer than the real diameter [23].

The 'largest-smallest diameter' method was used to measure the size of fibers in series of skeletal muscle biopsies of primary myopathies, infantile dermatomyositis, and secondary myopathies [4, 24]. Perimysial demarcation identified muscle fascicles, in which a peripheral subgroup of fibers was distinguished. In primary myopathies, including progressive muscular dystrophy and Becker-Kiener-type dystrophy, a subgroup of enlarged fibers is mixed with normal-sized ones, but with enlarged fibers mainly affecting the periphery of fascicles. In infantile dermatomyositis, muscle fiber diameter was significantly lower at the periphery.

In the paraspinous muscles in patients with idiopathic scoliosis, Slager and Hsu [25],

Table 1. Essential steps in research design for morphometry

Stage	Procedure	Common pitfalls
Population	Definition	Interference between variables used for definition and variables used for computation
	Randomization	Inadequate
	Fixation, processing	Inadequate handling of specimen
All the following	Definition of acceptable error (e.g. 5%) for sampling and computation	Error unsettled
Pilot cohort of samples	Prospective definition of dimension of adequate sampling	Inadequate sampling size
Final cohort of samples	Control of significance of sample dimension	Inadequate significance
Variables	Choice	Inappropriate for function being examined
	Anisotropism	No correction
	Significance of computation	Not significant
Derived functions	Compound significance	Not significant
Comparison with controls	Compound significance	Not significant

using the same method ('largest-smallest') found bilateral type 1 fiber atrophy at the apices of the curves and bilateral type 2 hypertrophy. These data suggest that muscle changes in idiopathic scoliosis are probably compensatory.

In an autopsy study of children aged 1–15 years, Pesonen et al. [26] used the ratio of the radius and the media thickness of the coronary arteries in semiserial cross sections as a measure of intimal thickness. Multifactorial analysis on the size of the arteries, the thickness of the arterial media and intima, and the cross-sectional areas of the arterial layers showed that media thickness of the anterior descending artery was the most dependent parameter, i.e., its changes were mostly accounted for by variations in the other parameters.

Pitfalls in Image Analysis

Classic morphometric methods are part of a large array of quantitative image analysis techniques. Their improper choice is the most common pitfall in quantitative analysis (table 1). Not only must the most appropriate technique match the problem, but the different sets of variables must also be identified and handled with adequate statistics [27, 28]. Overlooking some variables cannot be remedied at a later stage and may lead to wrong conclusions.

A practical example is found in electron microscopy. Variations in the amounts of submicroscopic structures may underlie important functional abnormalities. Visual recognition may be inadequate to recognize such variations, because of the smallness of the available sample. A quantitative approach may be informative. Given a measurement of variations in the thickness of the glomerular basement membrane (GBM), how can one conclude that the GBM in, for example, Alport's syndrome, is indeed thicker than normal, as suspected by visual examination?

Thickness of a biologic membrane can be derived from two classic morphometric functions, Vv and Sv [29]. First, ensure that the study and control populations are defined by parameters independent of those that will be quantified. Second, take into account the variation in size due to fixation and processing. The tissues of both the diseased and the control groups must be fixed and processed in the same way. Fixation shrinkage can sometimes be calculated by direct measurement. For instance, shrinkage of muscular fibers can be as high as 20% in formalin-fixed, paraffin-embedded material. Methacrylate embedding may result in less shrinkage [4]. In practice, these differences complicate comparisons of morphometric data from different studies using even slightly different procedures.

A more difficult issue is determining the sizes of the cohorts so that the figures are significant. Two stages in sampling affect significance of the calculation. How many glomerular sections should be examined in each kidney? How many kidneys should be included in a cohort? Several methods approximate the compound error due to the double set of samplings [7]. A statistical power analysis can help define the numbers necessary to show significance at predetermined intervals. In practice, reliability is served by taking a small number of measurements (glomeruli) from a large number of specimens (kidneys).

Both glomerular and renal samplings should be unbiased. A commonly encountered pitfall in sampling for morphometry, which is likely to be present in our example, is error due to anisotropism. Proper sampling presumes that all points are given the same chance of being chosen. If the structures being studied are distributed in space at random, the arrangement of points and lines in the sampling grid is inconsequential. On the contrary, if the structures have a nonrandom orientation (anisotropism), a bias may be introduced by the selection of the plane of section. Morphometric techniques are available to examine specimens from a varying point of view to minimize anisotropism. This issue is part of a general problem of quantitative analysis, i.e. defining and demarcating the reference space. Some classic morphometric variables, such as Sv, Lv, and Nv, not only rely on reference space for proper sampling, but also include a measure of it in their denominator. In other words, they relate the structure examined to space. Vv can be set apart from the above variables because it expresses a plain volume ratio, which does not include a measure of space.

Assuming the population is representative, an adequate number of measurements of the relevant variables should be taken. In particular, point-sampling methods should be employed with caution. If the component to be measured is small (e.g. its volume is less than 5% of the total), the values derived from point-counting may not be reliable. The size of the sample required for significance increases steadily with small components, according to Anderson and Dunnill's nomogram [7]. The tolerable error should be set in advance and checked during calculation. For the kidney, the error due to the calculation of GBM thickness depends on the compound error pertaining to the two parent variables, Vv and Sv.

Proper sampling is essential for accuracy, i.e. for enhancing closeness of estimate to the true value. Increasing the number of measurements in a biased specimen would merely increase their precision, i.e. reduce value dispersion. The use of automatic equipment for morphometry (planimeter) has little effect on accuracy. In contrast, precision of measurement is considerably enhanced, although the slower and more tedious traditional methods usually achieve an acceptable approximation.

Growth

Growth, which can be assessed quantitatively, may represent an important prognostic factor in many pathologic settings, such as tumors. The simplest method to discriminate tissues with different growth rates is counting mitotic figures; more refined methods for the recognition of proliferating cells include autoradiography and immunohistochemistry for markers of replication. Differences in growth rate between the various compartments of a structure underlie the morphologic changes during development [30].

Allometry is a method for computing relative growth. It relates a series of values representing a functional or morphologic variable to a paired series of values of another variable representing mass [31, 32]. The method, originally developed for comparative anatomy, has been applied to ascertain variation of a structure in series of different animal species (interspecific allometry) or in series of animals of the same species of different ages (intraspecific allometry). The equation of classic allometry,

$$y = aM^b,$$

which is usually applied in its logarithmic form,

$$\log y = \log a + b \log M,$$

relates y, the variable to be scaled, to M, a measure of mass, with b, the slope of the log-log plot, representing an index of relative growth of y on M.

If b equals the y/M dimensional ratio, then growth of y on M is isometric, i.e., the two series of values show consensual variations, with their relative proportions maintained. If b differs from the y/M dimensional ratio (positive or negative allometry), it represents a measure of their relative disproportionate growth. The standard deviation of b and the significance of the allometric equation can be calculated.

Allometry has been employed in biology, mainly in renal physiopathology, to relate systemic changes to kidney mass [33], and rarely to relate morphometric data in pathology [32].

Examples: The changes in the Vv and Sv of the surface and glandular epithelium of adenomatous polyps in a case of familial colonic polyposis were studied by the allometric equation [34]. The glandular epithelium grew at a faster rate than the surface epithelium, which grew isometrically with the polyps. As a result, increases in size of the polyps were mainly related to increases in the glandular component of the epithelium.

Variations in the Vv, Sv, and Lv of the fibrous septa of the liver in cases of cirrhosis pointed to a differential pattern of growth, identifying two subsets of cases characterized by either slender septa growing mainly longitudinally or short septa growing mainly in thickness [35].

Densitometry

Densitometry can be coupled to computer-assisted image analysis. A camera attached to the microscope analyzes the staining intensity of a given structure in comparison with a known standard. Instrumental limitations depend on the image resolution and the extent of the gray value scale. A comparison can be drawn with cytospectrometry [36], an older procedure whereby the light absorption of the microscopic structure is read stoichiometrically with a spectrometer. Although associated with higher resolution and precision, cytospectrometry is a tedious procedure that can be applied only to a limited number of specimens.

The most common application of densitometry is the assessment of nuclear DNA content as a function of the intensity of DNA staining (generally by the Feulgen technique) [37, 38]. The variables, besides the instrumental limitations, include reproducibility of the DNA stain and standardization of measurement with adequate controls. The results of densitometry are often complemented by the assessment of morphometric functions.

Examples: Nuclear DNA density, derived by computer-based image analysis of giant-cell granulomas of the jaws showed no significant differences between aggressive and nonaggressive types, both of which showed a diploid nuclear pattern [39]. In another study [40], aggressive giant-cell granulomas had a significantly higher density of giant cells and a higher surface area than the nonaggressive ones, but no differences in the number of giant-cell nuclei.

Computer-assisted densitometry has been occasionally applied to quantify the concentration of glomerular components. Aminopeptidase A was demonstrated histochemically in the glomeruli of mice and rats, and its concentration assessed by densitometry [41]. This experimental setting allowed quantification of aminopeptidase A activity in kinetic studies and following pharmacological manipulation.

Three-Dimensional Reconstruction

3DR benefits considerably from computer-assisted imaging, since traditional techniques for 3DR, which have been available for decades, have been applied only in few instances because they are difficult to reproduce and are time-consuming.

Example: The architecture of the palatine tonsil was reconstructed by tracing its surface contour in serial micrographs onto plastic film and by computerized image scanning [42]. This reconstruction showed, contrary to general views, that some germinal centers were located far from the tonsillar epithelium.

Computer-assisted image analysis may outline three-dimensional arrangement by combining reference traces of the profiles obtained in serial sections. With adequate computer memory, complete serial images of the whole structure can be stored. This complex procedure may be the forerunner of a diagnostic machine vision system [43].

Morphometry and Renal Disease

Introduction

Renal biopsies are currently interpreted through the subjective assessment of sections stained with traditional histochemical methods, with immunofluorescence and electron microscopy. The limited amount of tissue present in a biopsy and the limited spectrum of diagnostic patterns that are discernible histologically show the need for quantitation.

Unfortunately, many unsettled issues prevent the routine adoption of quantitative methods in the interpretation of renal biopsies, despite much recent work on renal morphometry. Disparate and often noncomparable methods make it difficult, if not impossible, to analyze results and generate reproducible criteria. In particular, reliable data on the range and tolerable variation of normal morphometric functions are an imperative prerequisite for defining renal disease. The complex architecture of the kidney poses peculiar problems which are not encountered in relatively simpler structures, e.g. skeletal muscle. An estimate of the total filtration surface must consider the number of the glomeruli, the independent filtration units of the nephrons, and the dimension of each of these units.

Decrease in Glomerular Mass and Renal Disease

Despite the poorly understood relationship between changes in glomerular mass and development of renal disease, several examples suggest an interplay between them. Oligomeganephronia is characterized by a congenital paucity of glomeruli, which are of increased size [44–46]. The mean area of the glomerular profiles in a case of unilateral oligomeganephronia with agenesis of the other kidney was 45,253 μm^2, contrasting with a range between 10,257 and 14,710 μm^2 in three age-matched controls. However, the Vv of the glomeruli was decreased due to a decrease in glomerular density in the renal cortex [46]. Oligomeganephronia inevitably progresses to glomerulosclerosis and renal insufficiency [47]. An increased risk for the development of glomerulosclerosis accompanies clinical and experimental conditions associated with a significant decrease in glomerular number, such as unilateral renal agenesis [48] and unilateral nephrectomy [49]. Strains of rats prone to develop hypertension have a reduced number of glomeruli as compared with normal or hypertensive-resistant strains [50, 51]. Following subtotal nephrectomy by ablation of one kidney and ligation of two out of three arteries of the contralateral kidney, rats develop hypertension and glomerulosclerosis [52].

In contrast to the above instances, in which the reduction in the number of glomeruli was marked, it is still unknown whether there is a relationship between mild to moderate reduction and the risk for development of hypertension and renal disease [51]. A physiologic increase in the interstitial Vv is associated with an increase in the number of obsolescent glomeruli

Table 2. Number of glomeruli per kidney in humans

Authors	Method	Patients, n	Glomeruli, n
Huschke [54]	Estimate	1	2.1×10^6
Kittelson [55]	Sections	1	1.04×10^6
Traut [56]	Estimate	23	$3.8–5.7 \times 10^6$
Vimtrup [57]	Maceration	5	$0.8–1.23 \times 10^6$
Moore [58]	Maceration	28	$0.6–1.2 \times 10^6$
Osathanondh and Potter [59]	Maceration	1	0.822×10^6
Elias and Hennig [60]	Sections	4	$1.3–1.7 \times 10^6$
Dunnill and Halley [61]	Sections	18	$0.6–1.5 \times 10^6$
Tryggvason and Kouvalainen [62]	Maceration	3	$1.35–2.3 \times 10^6$

after age 40 years [53]. The hypothesis [51] that only the subjects with a low number of glomeruli are susceptible to a chronic overload of sodium and develop hypertensive nephropathy has not been proven because there is no conclusive information on the range of variation of the glomerular number in the normal population.

Determining the Normal Glomerular Number

Two methods have been used to estimate the number of glomeruli present in the human kidney: (1) macerating the tissue and isolating glomeruli, or (2) counting glomerular profiles in histologic sections (table 2) [54–62].

As common bias of these studies, with some possible exceptions [56, 58], each of them only dealt with a few specimens. Traut [56] performed a rough estimation through indirect extrapolation of the total number from the number of glomeruli per single foramen papillarium. Moore [58] applied a sampling method and showed that the total glomerular number could be estimated within a range of 4% of the total glomerular count. He also found that the number of glomeruli in one kidney was within 10% of the number of glomeruli in the opposite kidney. Although Moore's patients included 19 whites and 9 blacks, his data were not broken down by race or body mass. Kittelson's [55] estimation of glomerular number, based on a meticulous reconstruction of serial sections, yielded values consistent with those of Moore's study, but was only applied on one specimen. Moore's results were also in good agreement with those of Vimtrup [57] and Osathanondh and Potter [59]. Using renal sections, Elias and Hennig [60] and Dunnill and Halley [61] also derived values which fell within the range counted by Moore [58].

The above studies employed traditional approaches to the determination of glomerular number that are cumbersome and can be applied only to a limited number of specimens. This is the main reason why, even if the accuracy of these methods appears to be good considering the reproducibility of the results from different authors, the data points are so few that it is impossible to subclassify glomerular number by age, sex, race or body mass index of the subjects. These issues need to be addressed because the above parameters may be crucial in view of some recent theories on the development of nephrosclerosis and hypertension that will be discussed in the following sections.

Applying new morphometric techniques, glomerular number can be evaluated more rapidly, and more specimens can be examined. An example is the fractionator method [63, 64], which can be useful for estimating the total number of glomeruli. With this method, particles are counted through exhaustive serial sections of small samples from the relevant organ. Accuracy is ensured by meticulous randomization of sampling and careful exclusion of additional variables that may bias the calculation.

Despite the above reservations, some of the traditional methods for counting glomeruli suggest that there may be a subset of 'normal' subjects with a low number. Moore [58] and Dunnill and Halley [61] identified a small group of subjects with values between 0.6×10^6 and 0.8×10^6, and hence considerably lower than those of the majority, which ranged between 0.8×10^6 and 1.2×10^6. This amount of variation requires far larger population samples to confirm the existence of a subset with low glomerular number. A corollary is the assessment of this subset in regard to sex, age, and race. If there were indeed a higher incidence of this trait in groups known to be at high risk for glomerulosclerosis, the theory of Brenner et al. [51] would be indirectly corroborated.

Determining Glomerular Size

Although the size of glomeruli in renal diseases has been evaluated more often than glomerular number, the figures published are difficult to compare because of their wide variation, a result in part of the different techniques employed. The problems of shrinkage and processing that were mentioned above may also play a role in this discrepancy. Most of the studies on glomerular size consider the glomerulus a sphere; this assumption is reasonable because the relationship between volume and mean axial values does not vary considerably between spheres and slightly prolate or oblate ellipsoids. However, as a common bias, these studies often calculate only the mean of a dimensional factor (e.g. diameter, area) of glomerular profiles. These measures relate to the central 'equatorial' glomerular section, but are affected by a number of spurious variables, such as section thickness, optically lost caps, or uneven size and shape of the glomeruli. More accurate methods are available [7], but have been only sparsely employed. They are based on the mean of central sections either measured directly or estimated from the integration of the values of glomerular profiles. Furthermore, methods available to identify size classes of glomeruli [65] could be applied to recognize a wider, more irregular distribution of glomerular sizes in diseases with normal mean glomerular areas.

In the pediatric age, defining the size of glomeruli is further complicated by their uneven development. The formation of nephrons lasts from the 9th to the 44th week after conception, with each newly formed nephron apposed to the periphery of each renal lobe. As a result, the earliest formed glomeruli are located centrally. Approximately 70–75% of the nephrons are found in the outer and midportions of the renal cortex [58]. The juxtamedullary glomeruli, being formed first, are more mature than the superficial ones during fetal development, and are bigger until early postnatal life. From the mean areas of glomerular sections intersecting the vascular pole in fetal and infantile kidneys, Souster and Emery [66] showed that glomerular size increases rapidly after birth, with the difference between juxtamedullary and superficial glomeruli tapering progressively and disappearing after age 3 years. Fetterman and Habib [44] found that the size difference between juxtamedullary and cortical glomeruli disappears by age 14 months. We believe that the discrepancies between the above authors could be due in part to their correlating glomerular size only with age without considering other variables. An essential baseline for studies on glomerular size in different areas of the kidney would be to ascertain whether the total filtrating surface of the glomeruli increases in parallel with body mass.

Another open issue is whether glomeruli at different cortical locations display a different susceptibility to disease. Juxtamedullary glomeruli are often spared in diffuse mesangial sclerosis, but are preferentially affected in the congenital nephrotic syndrome of Finnish type (CNS-F) [67, 68]. Possible morphologic differences between the juxtamedullary and the subcapsular glomeruli could underlie the differences outlined above, for instance through variations in the blood flow to which the glomeruli may be exposed. Such issues cannot be addressed in renal biopsies since the depth of the sample is seldom known.

Do juxtamedullary glomeruli constitute a subset of larger dimensions even in the adult human kidney? Olivetti et al. [69, 70] found a linear transcortical gradient of glomerular diameters in both young and old rats, with the mean volume of juxtamedullary renal glomeruli being twice that of subcapsular glomeruli. According to Arataki [71], the glomeruli of rats continue to increase in number and size until the age of 100 days. In man, no difference between juxtamedullary and subcapsular glomeruli after 14–36 months has been identified [55, 66]. Abrams and co-workers [72] measured the profile areas of glomeruli in samples taken from eight different sites of a kidney, and found no significant variation. However, they did not differentiate between superficial and deep glomeruli. Wiltrakis sampled two areas (polar and parahilar) of 12 normal and 14 abnormal human kidneys, and compared glomerular size in five different regions of these samples: juxtamedullary, central columns of Bertin, superficial, intermediate and deep cortical [73]. He found no significant difference in glomerular volume between these regions on an intrarenal basis. However, he found an inverse relationship between glomerular volume and kidney size in normal individuals, but a direct relationship between glomerular volume and kidney volume when comparing hypertrophic and atrophic kidneys. Many authors make a point of sampling both compartments (superficial and deep cortex) in an attempt to derive an average value. If a difference in size exists, each compartment should be sampled separately. If there is no difference in size between the compartments, the areas of profiles obtained through renal biopsies could be used to derive a reliable measure of glomerular size in that kidney. Corollary issues are determining (a) if only a subset of the general adult population shows a difference in size between juxtamedullary and subcortical glomeruli, and (b) if this hypothetical subset has a different incidence of nephropathy.

Relating Number, Size and Maturity to Disease

Although the differences in glomerular number and size appear to be important for the pathogenesis of glomerulosclerosis, the precise relationship among these three variables is unknown. First, number and size do not vary in parallel, e.g. some diseases have only a change in number, and others have only changes in size. Second, increases in size do not necessarily result in sclerosis. Third, the subtotal nephrectomy model (which is partly biased because animal kidneys often differ in their development from man's) shows that sclerosis may not be related directly to reduction in the filtering mass, but that it may be secondary to the concurrent hypertension. However, hypertension may not be the sole factor responsible for glomerulosclerosis.

There are elements to suggest that genetically determined developmental abnormalities may lead both to sclerosis and to variations in the number and size of the glomeruli, as well as in their maturation rate. A disturbance in the regulation of the development and the number of glomeruli has been implicated in the development of the renal lesions of CNS-F, an autosomal-recessive trait [74], which is associated with a significant increase in kidney weight [75, 76]. In CNS-F the number of nephrons is increased by 75% [62, 77] and the glomerular Vv is preserved, resulting in a greater absolute total glomerular mass [75].

A systematic quantitative approach could clarify the complex relationships regulating the number of glomeruli being formed during uterine life, their dimensions, and their rate of maturation. The anecdotal data on the quantitative changes in CNS-F mentioned in the previous paragraphs point to changes in the above factors being closely associated with the pathogenesis of this disease and possibly of the other congenital nephrotic syndromes of the first year of life.

After the first year of life, quantitative methods have proved important for establishing prognosis. In this respect, Fogo et al. [78] used morphometric techniques in a recent study to compare glomerular size between patients with minimal changes who progressed to focal and segmental sclerosis and patients with minimal changes who got better. They found that the patients who progressed had significantly larger glomeruli in the original biopsies showing only minimal changes histologically. In a similar study, Yoshikawa et al. [79] showed a wide variation in size of the glomeruli in patients with minimal-change disease, and consistently large glomeruli in focal glomerulosclerosis. However, no follow-up information on the patients with minimal change disease was provided.

Measuring Microscopic Components of the Glomerulus

The basement membrane surrounding the glomerular capillaries provides the only continuous barrier between plasma and renal filtrate. Qualitative alterations of the GBM have long been related to changes in filtration and to specific diseases of the kidney, e.g. Alport's syndrome. Setting a normal standard of GBM thickness and relating it to various disease states involves overcoming many of the technical problems mentioned in the first part of this article about possible applications of the method of Weibel and Knight [29]. The method of Jensen et al. [80] for membrane thickness determination has also been employed to measure the GBM (fig. 5). It is derived from Weibel and Knight's [29] method and relies on an integral

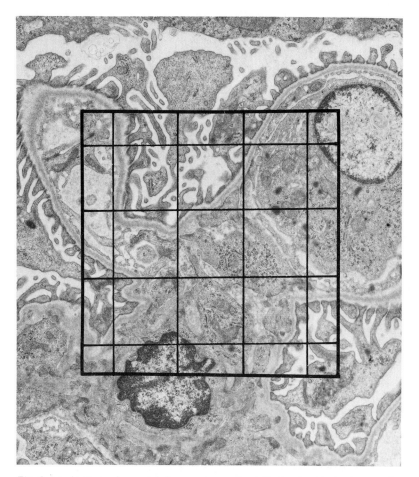

Fig. 5. An electron micrograph from a glomerular tuft showing the basic procedure for assessing GBM thickness with a superimposed grid of lines of known length [80]. Orthogonal intercepts are measured at the intersections between the lines and the inner border of the GBM, exclusive of the area adjacent to the mesangium.

equation relating the probability of orthogonal intercepts to thickness. Mauer et al. [81] applied this method to electron micrographs of human glomeruli from 118 normal subjects aged 9–65 years. The GBM thickness was measured in about 120 locations per biopsy. The Vvs of the glomerular components were derived from point counting, averaging about 3,000 points on 3.4 glomeruli per biopsy. These authors found that GBM thickness did not relate to albuminuria, hypertension or decreasing GFR in diabetic nephropathy. Jensen's method was also used by Dische et al. [82] to compare the thickness of the GBM in thin-membrane nephropathy with that of controls (296 vs. 400 nm). Steffes et al. [83] derived the thickness of the GBM from the harmonic mean of measurements taken on ×20,000 electron micrographs of normal kidneys from donors for transplantation. The mean thickness was 373 nm in males and 326 nm in females, with slightly higher values before the 4th decade of life. The volume of the mesangial component of the glomerular tuft was determined by Østerby [84] through planimetry on equatorial sections of the glomerulus derived from multiple electron micrographs. The mesangial volume did not change from the normal within 5 years from the onset of juvenile diabetes mellitus.

Perspectives for Quantitation in Renal Research

Current theories of glomerular hypertrophy hold that as some glomeruli become sclerotic, the remaining glomeruli enlarge with compensatory increase in filtering surfaces. However, there are examples in which this sequence seems not to occur. In CNS-F the glomeruli appear to be increased in number rather than in size. In the cases of minimal-change nephrotic syndrome that progress to focal and segmental sclerosis, the change in glomerular volume heralds sclerosis, suggesting a different pathogenesis between these two groups of nephrotic syndrome. This also seems to occur in diabetes mellitus type 1, a disease in which the glomeruli enlarge before becoming sclerotic.

In summary, quantitative methods may clarify the relationships among renal development, number and size of glomeruli, and the pathogenesis of glomerulosclerosis. The mechanisms regulating glomerular development and compensatory enlargement are unknown, and the connection between enlargement and sclerosis is not obligatory. Sclerosis may not be preceded by changes in the glomerular mass. Although glomerular enlargement has been described in several renal diseases that lead to glomerulosclerosis, such as diabetes mellitus type 1 and hypertension, it is not seen in diabetes mellitus type 2, a form that also evolves into sclerosis [81, 85, 86]. Incidentally, glomerular enlargement is associated with increased body mass in obese subjects [87], which again emphasizes the need to relate quantitative parameters in renal pathology to the anthropometric parameters of the patients. As already seen in CNS-F, increases in the number of glomeruli are not necessarily associated with increases in glomerular dimension.

Table 3. Kidney diseases in children and possible morphometric applications

Disease	Histologic appearances	Morphometry
Oligomeganephronia	Few glomeruli of increased size	Ascertain whether this represents the extreme end of a normal spectrum or a distinct entity
Congenital nephrotic syndrome of Finnish type (CNS-F)	Variable, ranging from minimal abnormalities to diffusely sclerotic glomeruli	Conflicting data suggest increase in the number of glomeruli and their maturation rate [76, 77]
Minimal-change disease	No light microscopic lesions; no immune deposits	A subset of patients with large glomeruli progress to focal glomerular sclerosis [80]
Focal segmental glomerulosclerosis	Areas of solidification in glomeruli and synechiae	Determine changes in the Vv of the glomerular compartments
Diabetes mellitus type 1	Increased mesangial matrix	Glomeruli are enlarged before becoming sclerotic [82–84]
Alport's syndrome	GBM becomes split and thickened	Identify mild ultrastructural changes in carriers
Thin-membrane disease	GBM becomes attenuated	Electron micrographs may differentiate from Alport's syndrome or IgA disease
IgA disease	Immune complex IgA deposits	Vv changes may define subsets of patients at risk

Perspectives for Quantitation in Renal Biopsy Interpretation

To improve the interpretation of renal biopsies, quantitation requires reliable values of normal and pathological ranges of variation. We expect that, with future advances in renal research, most of the above parameters could be measured in routine material. Table 3 shows some of the pediatric conditions the diagnosis of which could benefit from the application of morphometry. The mean size of the glomeruli could be estimated with a planimeter from the glomerular profiles, if their number is sufficient. The glomerular Vv in the cortex could likewise be measured. These data could lead to the estimation of the total number of glomeruli, if radiologic imaging could provide a sensible estimate of the total cortical mass. In addition, morphometric variables, such as Vv, Sv, and Lv, of the different glomerular compartments (e.g. GBM, mesangial matrix) could be calculated from electron micrographs with traditional methods or computer-assisted image analysis [86, 88, 89].

Currently, the presence and amount of abnormal glomerular components are appreciated subjectively and estimated on a poorly reproducible scale of 1+, 2+, 3+, and 4+. Densitometry by image analysis may represent a valuable tool in this respect. However, a number of obstacles must be overcome to standardize the measurement of glomerular sclerosis by densitometry, such as (a) setting normal and pathologic standards, (b) overcoming anisotropism of sclerosis distribution, (c) establishing a specific stain that could demonstrate sclerosis stoichiometrically, and (d) overcoming biases of sampling. Much more information can be derived from an objective approach to tissue diagnosis, but at a considerable cost (serial sections, tissue processing, standardization of staining, standardization of measurement, quality control programs). Such a goal can be reasonably achieved in the forthcoming future for only one or a few renal diseases, but the investment is justified because the information generated can enhance the accuracy of diagnosis and set new guidelines for prognosis. Once standards have been established for both procedures and equipment, densitometry will be available as a common diagnostic tool.

Absolute values, e.g. GBM thickness, could be estimated from the morphometric variables providing data essential to evaluate renal biopsies for congenital abnormalities, such as Alport's syndrome or 'thin-basement membrane' disease [90]. Densitometry could be employed to quantitate the amount of sclerosis as revealed with histochemical stains, such as the PAS reaction and silver impregnation, or, more precisely, with specific antibodies to different components of the intercellular matrix. The rate of growth of the glomerular cells could be assessed by autoradiography or with immunohistochemical methods for markers of nuclear replication, such as antibodies to cyclin and Ki-67.

References

1   Elias H: Identification of structure by the common-sense approach. J Microsc 1972; 95:59–68.
2   Weibel ER: Stereological methods in cell biology: Where are we – Where are we going? J Histochem Cytochem 1981;29:1043–1052.
3   Meltzer B, Searle N: Impotence principle in descriptive morphology. Nature 1968; 217:1289–1290.
4   Peiffer J, Bähr M: Anomalies in perifascicular muscle fibers as a differential-

diagnostic criterion. I. Perifascicular atrophy in inflammatory myopathies. Clin Neuropathol 1987;6:123–132.

5   Weibel ER, Elias H: Quantitative Methods in Morphology. Berlin, Springer, 1965.
6   DeHoff RT, Rhines FN: Quantitative Microscopy. New York, McGraw-Hill, 1968.
7   Aherne WA, Dunnill MS: Morphometry. London, Arnold, 1982.
8   Elias H. Hyde DM: A Guide to Practical Stereology. Basel, Karger, 1983.
9   Revell PA: Histomorphometry of bone. J Clin Pathol 1983;36:1323–1331.
10 Pesce C, Reale A: Testis morphometry in varicocele. Arch Androl 1985;15:193–197.
11 Meinhard EA, Warbrook DG, Risdon RA: Computer card morphometry of jejunal biopsies in childhood coeliac disease. J Clin Pathol 1975;28:85–93.
12 Sternlieb I, Feldmann G: Effects of anticopper therapy on hepatocellular mitochondria in patients with Wilson's disease. Gastroenterology 1976;71:457–461.
13 Pesce C, Colacino R: The sinusoids in cirrhosis. A morphometric study. Virchows Arch [A] 1986;410:217–219.
14 Pesce C, Colacino R: Morphometric analysis of capillary and cavernous hemangiomas. J Cutan Pathol 1986;13:216–221.
15 Fenton BM, Zweifach BW, Worthen DM: Quantitative morphometry of conjunctival microcirculation in diabetes mellitus. Microvasc Res 1979;18:153–166.
16 Lack EE, Perez-Atayde AR, Young JB: Carotid body hyperplasia in cystic fibrosis and cyanotic heart disease. A combined morphometric, ultrastructural, and biochemical study. Am J Pathol 1985;119:301–314.
17 Roberts PF: Variation in the morphometry of the normal human thyroid in growth and ageing. J Pathol 1974;112:161–168.
18 Witzleben CL, Finegold M, Piccoli DA, et al: Bile canalicular morphometry in arteriohepatic dysplasia. Hepatology 1987;7:1262–1266.
19 Seshadri R, Jarvis LR, Jamal O, et al: A morphometric classification of acute lymphoblastic leukemia in children. Med Pediatr Oncol 1985;13:214–220.
20 Tosi P, Luzi P, Miracco C, et al: Morphometry for the prognosis of acute lymphoblastic leukemia in childhood. Pathol Res Pract 1987;182:416–420.
21 Ouvrier RA, McLeod JG, Conchin TE: Morphometric studies of sural nerve in childhood. Muscle Nerve 1987;10:47–53.
22 Ouvrier RA, McLeod JG, Conchin TE: Friedreich's ataxia. Early detection and progression of peripheral nerve abnormalities. J Neurol Sci 1982;55:137–145.
23 Haug H: The significance of quantitative stereologic experimental procedures in pathology. Pathol Res Pract 1980;166:144–164.
24 Bähr M, Peiffer J: Anomalies in perifascicular muscle fibers as a differential-diagnostic criterion. Clin Neuropathol 1987;6:133–138.
25 Slager UT, Hsu JD: Morphometry and pathology of the paraspinous muscles in idiopathic scoliosis. Dev Med Child Neurol 1986;28:749–756.
26 Pesonen E, Hirvonen J, Laaksonen H, et al: Morphometry of coronary arteries: its use in children 1 year of age or older. Arch Pathol Lab Med 1982;106:381–384.
27 Pesce C: Defining and interpreting diseases through morphometry. Lab Invest 1987;56:568–575.
28 James NT: Common statistical errors in morphometry. Pathol Res Pract 1989;185:764–768.
29 Weibel ER, Knight BW: A morphometric study on the thickness of the pulmonary air-blood barrier. J Cell Biol 1964;21:367–384.

30 Meltzer B. Searle NH, Brown R: Numerical specification of biological form. Nature 1967;216:32–36.

31 Gould SJ: Allometry and size in ontogeny and phylogeny.Biol Rev 1966;41:587–640.

32 Pesce CM, Carli FS: Advances in formal pathogenesis: Allometry and allied comparative methods of stereology for recognizing relative growth in proliferative lesions. Pathol Immunopathol Res 1989;8:95–103.

33 Shea BT, Hammer RE, Brinster RL: Growth allometry of the organs in giant transgenic mice. Endocrinology 1987;121:1924–1930.

34 Pesce CM, Colacino R: Relative growth of adenomatous polyps of the colon. Stereology and allometry of multiple polyposis.Virchows Arch [A] 1987;412:151–154.

35 Pesce CM, Carli FS: Growth patterns of the fibrous septa of cirrhosis: A morphometric and allometric study. Pathol Res Pract 1989;184:486–488.

36 Auer G, Askensten U, Ahrens O: Cytophotometry. Hum Pathol 1989;20:518–527.

37 Bibbo M, Dytch HE, Bartels PH, et al: Clinical applications of a microcomputer-based DNA-cytometry system. Acta Cytol 1986;30:372–378.

38 Dytch HE, Bibbo M, Bartels PH, et al: An interactive microcomputer-based system for the quantitative analysis of stratified tissue sections. Anal Quant Cytol Histol 1987;9:69–78.

39 Eckhardt A, Pogrel MA, Kaban LB, et al: Central giant cell granulomas of the jaws. Nuclear DNA analysis using image cytometry. Int J Oral Maxillofac Surg 1989;18:3–6.

40 Ficarra G, Kaban LB, Hansen LS: Central giant cell lesions of the mandible and maxilla: A clinicopathologic and cytometric study. Oral Surg Oral Med Oral Pathol 1987;64:44–49.

41 Kugler P: Aminopeptidase A is angiotensinase A. I. Quantitative histochemical studies in the kidney glomerulus. Histochemistry 1982;74:229–245.

42 Abbey K, Kawabata I: Computerized three-dimensional reconstruction of the crypt system of the palatine tonsil. Acta Otolaryngol 1988;454:39–42.

43 Bartels PH, Thompson D, Bartels HG, et al: Machine vision system for diagnostic histopathology. Pathol Res Pract 1989;185:635–646.

44 Fetterman GH, Habib R: Congenital bilateral oligonephronic renal hypoplasia with hypertrophy of nephrons (oligomeganephronia). Am J Clin Pathol 1969;52:199–207.

45 Miltenyi M, Balogh L, Schmidt K, et al: A new variant of the acrorenal syndrome associated with bilateral oligomeganephronic hypoplasia. Eur J Pediatr 1984;142:40–43.

46 Nomura S, Osawa G: Focal glomerular sclerotic lesions in a patient with unilateral oligomeganephronia and agenesis of the contralateral kidney: a case report. Clin Nephrol 1990;33:7–11.

47 McGraw M, Poucell S, Sweet J, et al: The significance of focal segmental glomerulosclerosis in oligomeganephronia. Int J Pediatr Nephrol 1984;5:67–72.

48 Ashley DJB, Mostofi FK: Renal agenesis and dysgenesis. J Urol 1960;83:211–230.

49 Zucchelli P, Cagnoli L, Casanova S, et al: Focal glomerulosclerosis in patients with unilateral nephrectomy. Kidney Int 1983;24:649–655.

50 Brandis A, Bianchi G, Reale E, et al: Age-dependent glomerulosclerosis and proteinuria occurring in rats of the Milan normotensive strain and not in rats of the Milan hypertensive strain. Lab Invest 1986;55:234–243.

51 Brenner BM, Garcia DL, Anderson S: Glomeruli and blood pressure. Less of one, more the other? Am J Hypertens 1988;1:335–347.

52 Purkerson ML, Hoffsten PE, Klahr S: Pathogenesis of the glomerulopathy associated with renal infarction in rats. Kidney Int 1976;9:407–417.

53 Kappel B, Olsen S: Cortical interstitial tissue and sclerosed glomeruli in the normal human kidney, related to age and sex. A quantitative study. Virchows Arch [A] 1980; 387:271–277.

54 Huschke E: Über die Textur der Nieren. Oken's Isis (vol 21) 1828. (Cited after BJ Vimtrup [57].)

55 Kittelson JA: The postnatal growth of the kidney of the albino rat, with observations on an adult human kidney. Anat Rec 1917;13;385–408.

56 Traut HF: The structural unit of the human kidney. Contr Embryol 1923;15:103–120.

57 Vimtrup BJ: On the number, shape, structure, and surface area of the glomeruli in the kidneys of man and mammals. Am J Anat 1928;41:123–151.

58 Moore RA: The total number of glomeruli in the normal human kidney. Anat Rec 1931;48:153–178.

59 Osathanondh V, Potter E: Development of human kidney as shown by microdissection. IV. Development of tubular portions of nephrons. Arch Pathol 1966;82:391–402.

60 Elias H, Hennig A: Sterology of the human renal glomerulus; in Weibel ER, Elias H (eds): Quantitative Methods in Morphology. Berlin, Springer, 1967, pp 130–166.

61 Dunnill MA, Halley W: Some observations on the quantitative anatomy of the kidney. J Pathol 1972:110:113–121.

62 Tryggvason K, Kouvalainen K: Number of nephrons in normal human kidneys and kidneys of patients with the congenital nephrotic syndrome. Nephron 1975;15:62–68.

63 Gundersen HJG, Bendtsen TF, Korbo L, et al: Some new, simple and efficient stereological methods and their use in pathological research and diagnosis. APMIS 1988;96:379–394.

64 Gundersen HJG, Bagger P, Bendtsen TF, et al: The new stereological tools: Disector, fractionator, nucleator and point sampled intercepts and their use in pathological research and diagnosis. APMIS 1988;96:857–881.

65 Saltykov SA: The determination of the size distribution of particles in an opaque material from a measurement of the size distribution of their sections; in Proc 2nd Int Congr for Stereology, Chicago. Berlin, Springer, 1967, pp 163–173.

66 Souster LP, Emery JL: The sizes of renal glomeruli in fetuses and infants. J Anat 1980;130:595–602.

67 Hoyer JR, Vernier RL, Najarian JS, et al: Recurrence of idiopathic nephrotic syndrome after renal transplantation. Lancet 1972;ii:343–348.

68 Kaplan BS, Bureau MA, Drummon KN: The nephrotic syndrome in the first year of life: Is a pathologic classification possible? J Pediatr 1974;85:615–621.

69 Olivetti G, Anversa P, Rigamonti W, et al: Morphometry of the renal corpuscle during normal postnatal growth and compensatory hypertrophy. A light microscope study. J Cell Biol 1977;75:573–585.

70 Olivetti G, Anversa P, Melissari M, et al: Morphometry of the renal corpuscle during postnatal growth and compensatory hypertrophy. Kidney Int 1980;17:438–454.

71 Arataki M: On the postnatal growth of the kidney, with special reference to the number and size of the glomeruli (albino rat). Am J Anat 1926;36:399–436.

72 Abrams RL, Lipkin LE, Hennigar GR: A quantitative estimation of variation among human renal glomeruli. Lab Invest 1963;12:69–76.

73 Wiltrakis MG: Correspondence between glomerular size and renal volumes in atrophic, normal and hypertrophic human kidneys. Acta Pathol Microbiol Scand [A] 1972;80:801–811.

74 Norio R: The nephrotic syndrome and heredity. Hum Hered 1969;19:113–120.

75 Huttunen NP, Rapola J, Vilska J, et al: Renal pathology in congenital nephrotic syndrome of Finnish type: a quantitative light microscopic study on 50 patients. Int J Pediatr Nephrol 1980;1:10–16.

76 Giles HM, Pugh RCB, Darmady EM, et al: The nephrotic syndrome in early infancy: a report of three cases. Arch Dis Child 1957;32:167–180.

77 Tryggvason K: Morphometric studies on glomeruli in the congenital nephrotic syndrome. Nephron 1978;22:544–550.

78 Fogo A, Hawkins EP, Berry PL, et al: Glomerular hypertrophy in minimal change disease predicts subsequent progression to focal glomerular sclerosis. Kidney Int 1990;38:115–123.

79 Yoshikawa N, Cameron AH, White RHR: Glomerular morphometry. I. Nephrotic syndrome in childhood. Histopathology 1981;5:239–249.

80 Jensen EB, Gundersen HJG, Østerby R: Determination of membrane thickness distribution from orthogonal intercepts. J Microsc 1979;115:19–33.

81 Mauer SM, Steffes MW, Ellis EN, et al: Structural-functional relationships in diabetic nephropathy. J Clin Invest 1984;74:1143–1155.

82 Dische FE, Anderson VER, Keane SJ, et al: Incidence of thin membrane nephropathy: morphometric investigation of a population sample. J Clin Pathol 1990;43: 457–460.

83 Steffes MW, Barbosa J, Basgen LM, et al: Quantitative morphology of the human normal kidney. Lab Invest 1983;49:82–86.

84 Østerby R: A quantitative electron microscopic study of mesangial regions in glomeruli from patients with short-term juvenile diabetes mellitus. Lab Invest 1973; 29:99–110.

85 Butcher D, Kikkawa R, Klein L, et al: Size and weight of glomeruli isolated from human diabetic and nondiabetic kidneys. J Lab Clin Med 1977;80:544–553.

86 Ellis EN, Steffes MW, Goetz FC, et al: Glomerular filtration surface in type I diabetes mellitus. Kidney Int 1986;29:889–894.

87 Cohen AH: Massive obesity and the kidney. Am J Pathol 1975;81:117–129.

88 Wehner H: Quantitative Pathomorphologie des Glomerulum der menschlichen Niere. Stuttgart, Fischer, 1974.

89 Ellis EN, Mauer SM, Sutherland DER, et al: Glomerular capillary morphology in normal humans. Lab Invest 1989;60:231–236.

90 Basta-Jovanovic G, Venkataseshan VS, Gil J, et al: Morphometric analysis of glomerular basement membranes (GBM) in thin basement membrane disease (TBMD). Clin Nephrol 1990;33:110–114.

Karen Schmidt, MD, Department of Pathology, University of Illinois,
1601 Parkview Avenue, Rockford IL 61107–1897 (USA)

Subject Index